CURRENT AFRICAN ISSUES 60

I0027149

The Role of Food Banks in Food Security in Uganda

The Case of the Hunger Project Food Bank, Mbale Epicentre

Joseph Watuleke

NORDISKA AFRIKAINSTITUTET, UPPSALA 2015

Joseph Watuleke

INDEXING TERMS:
Food security
Food supply
Smallholders
Farming
Agricultural production
Food storage
Livelihood
Sustainable development
Uganda

The opinions expressed in this volume are those of the author
and do not necessarily reflect the views of the Nordic Africa Institute.

ISSN 0280-2171
ISBN 978-91-7106-761-6
Language editing: Peter Colenbrander
© The author and the Nordic Africa Institute
Production: Byrå4

Contents

Glossary

CIGI   Centre for International Governance
GDP   Gross Domestic Product
GOU   Government of Uganda
HLPE   High Level Panel of Experts
MoFPED   Ministry of Finance, Planning and Economic Development
MWLE   Ministry of Water, Land and Environment
NGO   Non-Governmental Organizations
NRDC   Natural Resource Defence Council
NZCCSS   New Zealand Council of Christian Social Services
UBOS   Uganda Bureau of Statistics
UCSCU   Uganda Savings and Credit Cooperation Union Ltd
UNCBD   United Nations Convention on Biological Diversity

Tables

Abstract

This study addresses the role the food bank plays in food security, sustainable livelihoods and building resilience to climate change among smallholder farmers in Uganda, and in particular eastern Uganda. Currently, it is difficult to measure the socioeconomic impact of the food bank on smallholder farmers in eastern Uganda due to the difficulty of isolating its contribution from that of interrelated programmes and farmer activities. It is, however, evident that the food bank plays a significant role in improving the smallholder farmers' food production and incomes. The food bank is actively engaged in training smallholder farmers in modern farming methods, providing improved seeds and safe storage facilities for farmers' produce, helping farmers to diversify their livelihood sources and providing climate-related information.

Prolonged drought and lack of access to sufficient seeds of good quality are the main sources of food insecurity among smallholder farmers. Distance from the food bank and lack of access to information are among the other factors that affected many farmers' ability to participate in food bank activities. Community ownership of the food bank is still lacking, and this is a long term threat to the sustainability of the project. There is therefore an urgent need to establish community-managed food banks at lower levels that ensure community ownership; equitably distribute benefits among target farmers; encourage seed-saving among farmers; initiate community-supported agriculture programmes to improve access to farm credit; and invest in rainwater harvesting for irrigation.

1. Introduction

Following the food price crisis of 2008, debates about global food security have increased (Wiggins 2008). Concerns about the impact of the crisis on the prospects for achieving the first Millennium Development Goal (MDG) "to end poverty and hunger" are also high (Ludi 2009). Research shows that soaring food prices mainly affect three groups: the poor whose ability to buy food is undermined; governments of low-income countries that face higher import bills, soaring costs for safety net programmes and political instability; and aid agencies that must cope with increased demands for food, cash and technical assistance (Wiggins 2008:1).

Most hunger is caused by a failure to gain access to locally available supplies of food or to the means to produce the food directly (Timmer *et al.* 1983:4). However, droughts in major wheat-producing countries in 2005-06; low grain reserves; high oil prices; a doubling of per-capita meat consumption in some developing countries and a diversion of 5 per cent of the world's cereals to agrofuels have been seen as the immediate factors leading to the food price crisis of 2008 (Shah 2008). The effects of food insecurity hit people with low or insecure access to food the hardest: the very poor; the landless and near-landless; and the disadvantaged – children, pregnant and lactating women and the elderly who have lost a productive role in their societies (Timmer *et al.* 1983).

Although food prices normalised after 2008, the United Nations Development Programme (UNDP) still warns that in the long term, climate change will undermine international efforts to combat poverty (UNDP 2008:1). It is also observed that climate change will steadily increase the exposure of poor and vulnerable households to climate-shock, thus placing increased pressure on their coping strategies and, over time, steadily eroding human capabilities (UNDP 2008:10). Two of the five mechanisms identified by UNDP through which climate change could stall and reverse human development directly affect food security. First, there is the effect on agricultural production and food security, through effects on rainfall, temperature and water availability for agriculture in vulnerable areas. Second, there is water stress and water insecurity due to changed run-off patterns and glacial melt, which will compromise flows of water for irrigation and human settlement (UNDP 2008:10).

In considering the climate change effect on agriculture, it should be remembered that the agriculture sector is the backbone of a majority of African economies (Ludi 2009:1). It is still the largest contributor to GDP, the biggest source of foreign exchange and the main generator of the continent's savings and tax revenue (NEPAD 2002:7). Agriculture employs over 80 per cent of Africa's workforce, and at the same time, farming and agribusiness together constitute nearly 50 per cent of Africa's economic activity (World Bank 2013a).

This implies that agricultural vulnerability to climate change will likely cripple the economies of a majority of African countries, including Uganda. Improved agricultural performance, on the other hand, has the potential to uplift the majority of the African population from poverty through increased rural incomes and purchasing power (NEPAD 2002:7).

Although it is the backbone of African economies, Africa's agriculture faces serious challenges, including land degradation, inadequate irrigation, rural-urban migration, political instability and stagnant economies (Dinar 2007). Agricultural production is sensitive to climate, and the effects of climate change on the continent may force large regions of marginal agriculture out of production by the end of this century (Dinar 2007; FAO 2003).

In 2003, the United Nation's Food and Agriculture Organisation (FAO) indicated there were many uncertainties as to when and where climate change would impact agriculture and food security (FAO 2003). The Intergovernmental Panel on Climate Change (IPCC), however, notes that multiple stresses such as limited water resources, loss of biodiversity, and air pollution have already increased sensitivity to climate change and reduced resilience in the agricultural sector (IPCC 2007:277). Given agriculture's salient role in Uganda's economy, it is clear that agricultural failure will emasculate the health and productivity of individuals, thus impeding social and economic development (Wahlberg 2008).

One astounding fact in the literature that motivated this study and shaped its subject matter is that, throughout much of the world, there is enough food produced to feed everyone (FAO 2011a; Leathers and Foster 2009). Yet 842 million people around the world go hungry (FAO, IFAD and WFP 2013). Each year, millions of tons of consumable surplus food and groceries are lost through waste. Much of the food grown, processed, produced and manufactured is never consumed due to "failure to harvest; post-harvest losses; product disposal due to expiration, overproduction, damage and market" (Klein 2013).

The United Nations World Food Programme (WFP) on 4 June 2009 reported that hunger kills more people every year than AIDS, malaria and tuberculosis combined; and that hunger (being underweight) was number one on the list of the world's top ten health risks, while one in seven people (adults and children) go to bed hungry each night (WFP 2009). Why is this so? How can this situation be reversed? What is already being done about it? How can the food bank help? To these and many more questions everyone around the world would love to see answers.

Research on food banks and food security has focused on understanding why the number of food bank users is on the increase, and on improving diets at food banks by serving nutritious foods (O'Brien 2004; Moldofsky 2000; Handforth et al. 2013; McPherson 2006). However, none of these studies establish how food bank users could be increased and their ability to meet their food requirements in a sustainable manner improved. Moreover, most of these re-searches have been conducted in developed countries, whose context is different from that of developing countries. Thus, there is a need for further research into the role of food banks in developing countries like Uganda.

A number of studies also suggest that the problem of food insecurity can be reduced if investment in smallholder farmers is increased. Given that the food bank in Uganda is actively engaged with smallholder farmers in rural areas, there is a need to establish how it supports food security. The current study investigates issues related to access to financial capital, human capital, physical capital as well as social capital in pursuance of food security and sustainable live-lihoods among smallholder farmers. It also explores the sources of smallholder farmers' food security and livelihoods as well as their adaptability to climate variability.

Geographically, Uganda is a landlocked East African country. It lies astride the Equator at latitude 4° 12'N and 1° 29'S and, longitude 29° 34'E and 35° 0'W (Uganda Bureau of Statistics 2002). It is bordered by the Democratic Republic of Congo to the west, Kenya to the east, Rwanda and Tanzania to the south, and South Sudan to the north. Uganda is divided into three main environmental/geographical areas: swampy lowlands, a fertile plateau with wooded hills and a desert region (Game Plan Africa 2012).

Uganda occupies an area of 241,038 sq. kms, of which 43,941 sq. kms are open water and swamps, and 197,097 sq. kms is land (Uganda Bureau of Statistics 2002:1). The Mbale district, where the study was conducted, occupies 0.27 per cent (534.4 sq. kms) of this total land area. The district lies on the fertile plateau below the slopes of Mount Elgon (Uganda Bureau of Statistics 2002).

Socioeconomically, Uganda is a developing country with considerable natural resources, including fertile soils, regular rainfall, small deposits of copper, gold and other minerals and recently discovered oil. Uganda's GDP by the end of 2012, according to the World Bank, was estimated at $51.27 billion, with a real GDP growth rate of 2.6 per cent (World Bank 2013a). Agriculture is the most important sector in Uganda's economy, employing over 80 per cent of the workforce (MoFPED 2012). In Uganda, agriculture mainly comprises crop production, livestock production, fisheries and forestry. This study concentrated on farmers involved in crop production and mixed crop and animal production.

Common crops include coffee, beans, plantain (matooke), maize, onions, carrots, cassava and Irish and sweet potatoes (MoFPED 2012; Ssewanyana and Kasirye 2010). Over 80 per cent of Uganda's population is involved in subsistence agriculture, commonly referred to as smallholder farming. They rely on family labour and on simple farming tools and methods. They also face challenges in gaining access to enough good quality seed and storage facilities; in adapting to climate variability; and in the form of low yields, given limited use of fertiliser (Kasule *et al.* 2011).

The population of Uganda was estimated at about 36 million in 2012 (World Bank 2013a) with a population growth rate of 3.32 per cent, one of the highest in the world. Of this total, the estimated population of Mbale district was 410,300, approximately 1.14 per cent of the country's total population.

The highest percentage of the country's population, 48.9 per cent (male 8,467,172, female 8,519,723) is below the age of 15; and 4.4 per cent of the population is 55 and above (UBOS 2010). That means the country's productive population, between the ages of 15 and 54, is only 46.7 per cent. This signifies a huge dependent population (53.3 per cent), with significant implications for the country's economy (UBOS 2006). About 85 per cent of Uganda's total population is rural based and engaged in smallholder agriculture. In Mbale district, 92 per cent of the population is found in rural areas (UBOS 2010:6).

Ethnically, Ugandans can be classified into several broad linguistic groups: the Bantu-speaking majority, living in central, southern, western and some parts of eastern Uganda; and non-Bantu speakers, who occupy most of the eastern, northern and northwestern portions of the country. The latter can in turn be subdivided into Nilotic and Central Sudanic peoples.

The Bantu category includes the large and historically highly centralised kingdom of Buganda; the smaller western Ugandan kingdoms of Bunyoro, Nkore and Toro; and the Busoga states and Bugisu to the east of Buganda. The peoples in the second category include the Iteso, Langi, Acholi, Alur, Karamojong, Jie, Madi and Lugbara in the north, and a number of smaller groups in the eastern part of the country (Nyeko 1996).

The Baganda in the central region comprise the largest ethnic group (16.9 per cent of the total population), followed by the Banyakole (9.5 per cent), Basoga (8.4 per cent), Bakiga (6.9 per cent), Iteso (6.4 per cent), Langi (6.1 per cent), Acholi (4.7 per cent), Bagisu (4.6 per cent), Lugubara (4.2 per cent), Bunyoro (2.7 per cent) and other (29.6 per cent) (UBOS 2002). The original inhabitants of Mbale district are the Bagisu, part of the Bantu-speaking group.

The Bagisu speak Lu'Masaba, also known as 'Lugishu' or 'Gishu'. The author is a native speaker of Gishu. However, English is the official and uniting language and Luganda is another commonly spoken language in Uganda. Swahili is also being promoted in the spirit of regional socioeconomic integration into the East African Community and it is hoped that soon Swahili will be used as Uganda's national language. This research was conducted in the Bagisu ethnic region in Mbale district. The majority of the participants were Bagisu, except for a few of the Hunger Project staff.

Overview of the Hunger Project and food bank

The Hunger Project was established in 1977 to generate in a global context the will and commitment to end hunger on the planet by the end of the century (Lofchie and Commins 1984; Susan 1987). It is currently active in South Asia, sub-Saharan Africa and Latin America, where the highest concentrations of hungry people live (Hunger Project 2013a). It provides the tools and training to increase farm production and income-generating activities at the local level; empowers partners to create, stock and manage their own food banks; and encourages clusters of rural villages to develop sustainable, self-reliant, hunger-free communities. The income-generating activities enable women and men in participating communities to increase their incomes so they can purchase the food they need.

The Hunger Project also promotes sustainable farming practices: for example, local agricultural experts teach Hunger Project partners how to establish and manage community farms. In this programme, villagers learn techniques to improve crop yields sustainably and provide entire communities with increased access to food. The project also runs a microfinance programme, which trains and empowers villagers, with a special focus on women food farmers, who grow 80 per cent of household food in Uganda. In this programme, partners learn how to increase their incomes and use their savings to improve the health, education and nutrition of their families.

This research focused on the food security aspect of the Hunger Project, and concentrated on the role of food banks, using Mbale epicentre food bank as its case study.

3. Sustainability and sustainable development

Since sustainable development first appeared in the 1987 Brundtland Report by the World Commission on Environment and Development (WCED), the concept of sustainability has provoked numerous debates. Questions have emerged about what should be regarded as sustainable development and what not. It is therefore important to look more closely at the concept of sustainability.

Of course, like other complex subjects, sustainability does not have a universal definition. Some organisations, like the US Environmental Protection Agency (EPA), derive their explanation from the WCED definition of sustainable development. To them, sustainability is based on a simple principle: "everything that we need for our survival and well-being depends, either directly or indirectly, on our environment" (EPA n.d.). The agency therefore argues that sustainability creates and maintains the conditions under which humans and nature can exist in productive harmony, and permits the fulfilment of the social, economic and other requirements of present and future generations.

At the United Nations Earth Summit in 1992 in Rio, representatives of governments, the private sector and civil society addressed such themes as "how to build a green economy to achieve sustainable development and lift people out of poverty" and "how to improve international coordination for sustainable development" (ISGN Insights 2012). These deliberations resulted in Agenda 21, the Rio Declaration on Environment and Development, the Statement of Forest Principles, the United Nations Framework Convention on Climate Change and the United Nations Convention on Biological Diversity (UN 1997).

Since then, several agreements have been signed at international summits setting targets for the progressive achievement of the sustainability goal. These include: the Commission on Sustainable Development (December 1992); Millennium Development Goals, agreed at the Millennium Summit (2000); the Johannesburg Plan of Implementation, the World Summit on Sustainable Development (WSSD) in 2002; Rio Convention on Biological Diversity (UN-CBD); Kyoto Protocol to the UN Framework Convention on Climate Change (UNFCCC) in 1998 (see Hannah and Dubey n.d.; UN Kyoto Protocol 1998; UN Sustainable Development platform 2011.).

These targets and goals shaped the UN's understanding of sustainability. Thus, for the UN sustainability means a decent standard of living for everyone today without compromising the needs of future generations (UN Rio+20 2013). The realisation of these targets and goals, however, calls for identification of appropriate ways to help the poor climb out of poverty and get decent jobs without harming the environment.

The main message in the Global Environmental Outlook-5 (GEO5) chapter 16, "Scenarios and sustainability transformation," is that meeting an ambitious set of sustainability targets by the middle of the century is possible. The challenge, however, is the lack of adequate supporting policies and strategies to achieve it (GEO5 2012:420-2). However, one may ask how easy it will be to achieve sustainable development amidst great poverty in many developing countries, and in circumstances of global food insecurity, with about 842 million people around the world currently suffering from chronic hunger (FAO, IFAD and WFP 2013).

As we strive to achieve sustainable development, we need to consider that the majority of people in developing countries are still poor and engaged in subsistence agriculture using simple techniques. How the livelihoods of such people can be improved and sustained at a time when the world's food security is at a crossroads, is a question researchers need to explore.

World food security at a crossroads

The world is experiencing rising demands for food, stemming from three key forces: increasing human population; rising meat and dairy consumption with growing affluence; and biofuel consumption (Ray *et al.* 2013). The cost of food imports and factory farming are increasing and world food security is at a crossroads. In October 2009, during a High-Level Expert Forum in Rome, FAO predicted that the world population by the year 2050 would reach 9.1 billion, 34 per cent higher than it was in 2009. This would necessitate increased food production (net of food used for biofuel) by 70 per cent. Annual cereal production will need to rise to about 3 billion tons from 2.1 billion produced in 2009 (FAO 2009), and global food demand will increase by 60 per cent (IFAD, WFP and FAO 2012:30). This will be a challenge, especially in sub-Saharan Africa and in Uganda in particular, a country experiencing very high population growth rates, given the severe impacts of climate change, land and water degradation as well as the burden of HIV/AIDS.

Production is not meeting increasing food demand because, globally, food insecurity is largely a problem of access to the resources or services needed by families to produce, purchase or otherwise obtain enough nutritious food (FAO 2014a). The possibilities for increasing food production seem inadequate, owing to the fact that the natural resource factors on which agriculture depends have degenerated faster in the past 50 years than ever before (Neely and Fynn n.d.:5). This situation has divided the world into the "haves" and the "have nots." While poor countries suffer from a lack of food and malnutrition, obesity is a challenge for high and middle-income countries, accounting for 2.8 million adult deaths each year (UN 2013). At the same time, one-third of food produced for human consumption is wasted (Ken 2013). FAO researchers established that every year consumers in rich countries waste about 222 million metric tonnes of food, almost equivalent to the net food production of sub-Saharan Africa. Of these, fruits and vegetables, roots and tubers have the highest wastage rates (FAO 2014b).

In financial terms, food losses and waste amounts to roughly US$ 680 billion in industrialised countries and US$ 310 billion in developing countries (FAO 2014b). FAO also indicates that per capita food waste by consumers is between 95-115 kg. a year in Europe and North America, while each consumer in sub-Saharan Africa, South and Southeastern Asia throws away only 6-11 kg. a year.

Food waste in developing countries generally results from premature harvesting; poor post-harvest handling due to poor storage facilities and infrastructure; lack of processing facilities; and inadequate market systems. Generally, food waste by consumers is minimal in developing countries (FAO 2011a). On the other hand, food waste in industrialised countries results when production exceeds demand and from high "appearance quality standards" in supermarkets for fresh products; food deemed not fit for human consumption resulting from failure to comply with food safety standards; a "disposing is cheaper than using or re-using" attitude; large volumes on display and a wide range of products/brands; and consumer attitudes (FAO 2011a).

The concept of food insecurity and food security, however, is not new to the public – it has historical origins. Proper intervention in a country's food security matters requires a clear understanding of the historical perspectives, global and national concerns, as well as schools of thought on increasing food production.

Concept of food security

Food security is being widely discussed in global forums. Although it affects almost everyone on the globe, sub-Saharan Africa has widespread and chronic food insecurity. As of May 2006, for example, of 39 countries in the world experiencing serious food emergencies and requiring external assistance, 25 were in Africa, 11 in Asia and the Near East, two in Latin America and one in Europe (see Table 1).

FAO has observed that the number of food emergencies has risen from an average of 15 per year in the 1980s to more than 30 per year from 2000 onwards (FAO 2006). Major human-induced food emergencies persisting over several

Table 1: World food emergencies, 2005

Dominant variable	Africa	Asia	Latin America	Europe	Total
Human	10	3	1	1	15
Natural	8	7	1	0	16
Combined	7	1	0	0	8
Total	25	11	2	1	39

Source: FAO (2006)

years are known as protracted emergencies. These crises affect Africa more than any other region. In Africa, the average number of crises has tripled in the last two and a half decades. Why is this? What has been done? And what can be done?

FAO (2006) established that these food crises are mainly fuelled by armed conflict, often compounded by drought, floods and the effects of the AIDS pandemic. Conflict has a vast impact on food production and food security as millions of people are driven from their homes and unable to work their fields; they are also cut off from markets for their produce and from commercial supplies of seed, fertiliser and credit.

This study brings together the perspective of food banking and of food security in the Ugandan context. It argues that food security in Uganda is not a question of limited food production. It is rather a question of guaranteeing the ability to purchase available food on the market.

Historical perspective on food security

The question of food security has had a serious impact on the debate on global development since the 1970s, and has been in the public eye for a long time. The historical perspective on food security can be traced all the way to the Genesis, Chapter 41, verses 1–41. In interpreting Pharaoh's two dreams, Joseph predicted seven years of abundant harvest to be followed by seven years of famine. He thus advised Pharaoh to save food. People all over Egypt were asked to save a fifth of their grain harvest in the grain stores established in their respective cities. In this story, we see planning for food security, as well as food banking.

Historically, grain has been the principal food stored to ensure food security, and this is true today as well (Hunger Project 1985:94). However, the system by which grain is bought, sold and stored today is subject to severe fluctuations, especially when major producers like the US and the former USSR have bad harvests. This was evident in 1972–74 and 1979, when the USSR reduced production (Hunger Project 1985), and the 2005–07 production failures in major producing countries due to bad weather (OECD 2008). These resulted in worldwide food price increases, and many people in Third World countries suffered greatly.

Food security in global perspective

At a global level, discussion about food security and responses to it can be traced back to the 1943 Hot Springs Conference on Food and Agriculture, the establishment of FAO in 1945 and the Universal Declaration of Human Rights in 1948, which made access to adequate food a human right. Several global initiatives related to food security have since been pursued, as shown in Table 2. All of them aimed at finding a global solution to hunger and creating food security.

Table 2: Initiatives related to food security, 1943–92 (and 2014)

Year	Initiative
1943	Hot Springs Conference on Food and Agriculture is convened, at which freedom from want in relation to food and agriculture is defined as "a secure, an adequate and a suitable supply of food for every man"
1944	The International Bank for Reconstruction and Development (World Bank) and the International Monetary Fund (IMF) established
1945	FAO is founded in Rome
1946	United Nations International Children's Emergency Fund (UNICEF) and General Agreement on Tariffs and Trade (GATT) established.
1948	Freedom from Hunger and Malnutrition are recognised as a basic human right in the Universal Declaration of Human Rights
1951	Canada provides first bilateral food aid to India
1954	The US Agriculture, Trade, Development and Assistance Act 1954 (Food for Peace) (P.L.480) is signed, which allows the selling and bartering of surplus agricultural commodities for overseas development.
1954	The FAO Consultative Subcommittee on Surplus Disposal (CSD) is established to examine and regulate the impact of surplus disposal programmes.
1963	The World Food Programme is set up by the UN and FAO to use food aid for economic and social development, as well as emergency relief.
1965	The United Nations Development Programme is established.
1966	A Development Assistance Committee (DAC) high-level meeting adopts a recommendation on food problems in less-developed countries, stressing the need for higher food production and increased capital and technical assistance to support effective domestic agricultural policies.
1967	The International Grains Arrangement includes a Food Aid Convention (FAC) of 4.2 million tons of cereal aid.
1971	A Consultative Group on the International Agricultural Research (CGIAR) is established with sponsorship from the World Bank, FAO and UNDP
1972	The world food crisis marks the transition from an era of abundant supplies of cheap food for export and excess production capacity to one of highly unstable food supplies and prices.
1974	World Food Conference convenes in Rome. This is the first UN ministerial-level meeting on world food problems since the Hot Springs Conference of 1943. Food security is defined as ensuring the physical availability of food supplies in the event of widespread crop failure.
1974	World Food Council (WFC) established
1975	FAO's Global Information and Early Warning System (GIEWS) established
1975	International Emergency Food Reverse (IEFR) established with a minimum annual reserve of 500,000 tons
1975	Food and Nutrition Surveillance (FNS) activities initiated
1976	Food Security Assistance Scheme (FSAS) established
1977	International Fund for Agriculture Development (IFAD) is established by Canada
1979	WFC endorses a new Food Strategy Approach at the national level. Canadian International Development Agency (CIDA) commits $1 million to National Food Strategies but spends less than $100,000
1980	Food Aid Convention (FAC) is enlarged to 7.6 million tons of cereal aid annually
1981	The Pisani Memorandum outlines the European Community's commitment to National Food Strategies
1981	The IMF Compensatory Financing Facility is enlarged to assist countries encountering balance of payments difficulties arising from rising costs of cereal imports
1984	FAO endorses a broader definition of food security to ensure that all people at all times have both physical and economic access to the basic food they need.
1984	Canada announces the creation of a Special Fund for Africa to address drought and famine
1985	United States Department of Agriculture Economic Research Services releases the first World Food Needs and Availability report
1985	The Compact on Food Security receives support from the majority of members during the 23rd session of the FAO conference
1986	World Bank publishes "Poverty and Hunger: Issues and Options for Food Security in Developing Countries," in which food security is defined as access by all people at all times to enough food for an active, healthy life.

1987	The Inter-agency Food and Nutrition Surveillance Programme (IFNS) is initiated jointly by FAO, WHO and UNICEF
1987	WFP establishes the International Food Aid Information System (INTERFAIS) to assist in international coordination of food aid operations and policies
1987	USAID's Development Fund for Africa (DFA) is established and improving food security is identified as one of four key strategic objectives. In addition, USAID initiates Famine Early Warning Systems (FEWS) in eight African countries
1987	The first meeting of the WFC/UNICEF/ILO Consultation on the Impact of Economic Adjustment on People's Food Security and Nutritional Levels in Developing Countries held in Rome
1988	World Bank establishes a Food Security Unit within the Africa Regional Technical Department
1990	UN General Assembly unanimously adopts the International Development Strategy for 1990s, whose first priority is the eradication of poverty and hunger to be achieved through the implementation of four hunger-alleviation goals.
1990	The US Food, Agriculture, Conservation and Trade Act of 1990 (P.L. 101-624) is signed. It states that the primary motive for the use of food aid is food security.
1991	The Food Security Unit of the World Bank completes eight food security action plans. The FAO FSAS completes Phase I of four food security country studies
1992/ 2014	The first joint FAO/WHO international conference on nutrition is planned for December 1992 and improving household food security is identified as a central theme of the conference. The second International Conference on Nutrition (ICN2) is planned for 19-21 November 2014 as a positive, proactive global policy response to unacceptably high and persistent levels of malnutrition

Source: Adapted from Phillips et al. (1991) and FAO (2014c)

Because man needs food every day, yet the harvest is only at infrequent times each year, increased production and effective food storage have been considered critical to any programme to end hunger (Hunger Project 1985).

Increasing food production

Many ideas have been advanced to explain how global food production can be increased using basic resources to grow more food efficiently. Scientists such as Norman Borlaug advocated the adoption of the "Green Revolution" for increasing food production (Brown 1970; Borlaug *et al.* 1969). The Green Revolution involves the successful introduction of newly developed high-yielding varieties of grain (wheat, rice and others) in Third World countries (Hunger Project 1985:110). Borlaug in 1970 received the Nobel Peace Prize for his work in breeding the first high-yielding wheat varieties (Nobel Peace Prize 1970). Today, the term "Green Revolution" refers to almost any package of modern agricultural technologies introduced in the Third World.

This approach has, however, created a number controversies, with sceptics such as Raj Patel (2011) seeing it as a cause of social upheaval in peasant culture. Critics argue that the Green Revolution has not only failed to improve the lot of the poor, but has also led to ecological problems. The Green Revolution involves the use of modern technology, which the poor cannot afford, although poor farmers could eventually catch up with the larger farmers (Eicher 1995; Herath and Jayasuriya 1996). Lipton recognises that poverty delays the adoption of technology by smallholder farmers (Lipton 1989:118). Thus, whereas some people recommend the Green Revolution to increase food production, others advocate appropriate organic technologies (Hunger Project 1985).

Schools of thought on food production

Two main schools of thought on food production have been popular since the 1940s (Hunger Programme 1985:112–20). The first proposes that modern technologies offer an effective way to end hunger. Several points of view are evident in this school. They include a) Modern technology is the best method of food production; b) Science and technology offer particular advantages in terms of agricultural modernisation; c) New technologies can promote positive social and political change; and d) Technology can have a beneficial effect on the environment (Hunger Programme 1985).

The second school argues the contrary. It agrees that food production is a key element in ending hunger. However, it promotes different agricultural methods. For example, it advocates more organic methods of production, ones that do not depend on the intensive use of energy, chemicals or pesticides. The proponents of this alternative school contend that this approach has the merit of being ecologically sound, sustainable over a long period and the potential to be as productive as more mechanised forms of farming (Hunger Programme 1985).

In my view, as much as there has been increased adoption of agricultural technology in Uganda, given the poverty of many smallholder farmers in the country, the adoption of the ideas of second school will be more efficient. Farmers would only need to be trained in the use of locally available resources to boost production. For example, this study established that many smallholders could not afford pesticides or artificial fertilisers. However, they were mixing shrubs, hot pepper and animal urine for use as pesticides, and this practice was working for them. Others were using organic and composted manure on their gardens. This is particularly important, given changing trends in the understanding of food security, with the focus zooming in on individuals and households as the key units of analysis.

Changing trends in food security

It should be acknowledged that the subject of food security has kept changing in the past as a result of the emergence of global development as well as the dynamic nature of food problems around the world. Even the thinking on food security has gradually shifted from global and national food supplies to concerns about household and individual access to food (Devereux 2001). There is, however, the swinging pendulum between food supply and food consumption, implying a debate on whether the main focus of food security should be food production and supply or accessibility to food and consumption. Some of the details will be discussed in the next sections, where we look at the non-agricultural population and its food consumption. This has led to what I call an evolution of food security concerns.

Evolution of food security concerns

As the understanding of food security has evolved, so has its definition by different authors. For example, the definition derived from the World Bank conference of 1974 laid emphasis on food supply and focused mainly on food availability and stable food prices. Since then, however, the definition has shifted to include multidimensional concepts such as food accessibility, food utilisation and food stability; as well as bringing in the importance of households and individuals in food security concerns. Writers such as Sen (1981) have dominated this debate, raising the issue of entitlement. The debate has resulted in the shift from global and national concerns to individuals and households. In 1983, FAO modified its definition of food security thus: "Ensuring that all people at all times have both physical and economic access to the basic food they need" (FAO 1983).

From this definition, two important concepts can be derived: first, sustainable food production and second, sustainable livelihoods that enable people to access food at all times. Maxwell presents these distinctive variables clearly in his 1988 definition of food security, in terms of which a people or country are food secure when there is an efficient food system that removes the fear there will not be enough to eat (Maxwell 1988). Clearly, the understanding of food security and its definition have evolved a great deal since the 1974 World Food Conference. Some of the definitions of food security and food insecurity are presented in Box 1.

To bridge the different dimensions in understanding food security, Maxwell (1998) identifies three main shifts in food security trends. These include a shift from the global and national to the household and individual; from a food first perspective to a livelihood perspective; and from objective indicators to subjective perspectives. He establishes that there is consistency between these shifts and postmodern thinking in other spheres. For purposes of this study, only two of these spheres will be elaborated; that is, from the global and national to the household and individual; and from a food first to a livelihood perspective.

Shift from the global and national to household and individual

In this context, Maxwell suggests that the focus needs to shift from supply, as reflected in the 1975 UN definition, to concerns about national self-sufficiency. It can, however, be argued that widespread hunger can coexist with adequate food supplies at both national and international levels (Devereux 2001). This means that the argument that food security results from poor food supply might

Box 1: Definitions of food security, 1975–96

Author	Definition
United Nations 1975	"Availability at all times of adequate world supplies of basic food-stuffs … to sustain a steady expansion of food consumption … and to offset fluctuations in production and prices."
Reutlinger and Knapp 1980.	"A condition in which the probability of a country's citizens falling below a minimal level of food consumption is low."
Siamwalla and Valdes 1980	"The ability to meet target levels of consumption on a yearly basis."
Kracht 1981	"Everyone has enough to eat at any time –enough for life, health and growth of the young, and for productive effort."
Valdes and Konandreas 1981	"The certain ability to finance needed imports to meet immediate targets for consumption levels."
Reutlinger 1982	"Freedom from food deprivation for all of the world's people all of the time."
FAO 1983	"Ensuring that all people at all times have both physical and economic access to the basic food they need."
Heald and Lipton 1984	"The stabilization of access, or of proportionate shortfalls in access, to calories by a population."
Oshaug 1985, in Eide et al. 1985	"A basket of food, nutritionally adequate, culturally acceptable, procured in keeping with human dignity and enduring over time."
Reutlinger 1985	"Access by all people at all times to enough food for an active and healthy life."
World Bank 1986	"Access by all people at all times to enough food for an active, healthy life"
Zipperer 1987	"Always having enough to eat."
Barraclough and Utting 1987	"An assured supply and distribution of food for all social groups and individuals adequate in quality and quantity to meet their nutritional needs."
Falcon et al. 1987	"Both physical and economic access to food for all citizens over both the short and the long run."
Maxwell 1988	"A country and people are food secure when their food system operates efficiently in such a way as to remove the fear that there will not be enough to eat."
UN World Food Council 1988	"Adequate food available to all people on a regular basis."
Sahn 1989	"Adequate access to enough food to supply energy needed for all family members to live healthy, active and productive lives."
Reardon and Matlon 1989	"Consumption of less than 80% of WHO average daily caloric intake."
Sarris 1989	"The ability … to satisfy adequately food consumption needs for a normal and healthy life at all times."
Eide 1990	"Access to adequate food by and for households over time."
Phillips and Taylor 1990	"Food insecurity exists when members of a household have an inadequate diet for part or all of the year or face the possibility of an inadequate diet in the future."
Staatz 1990	"The ability … to assure, on a long-term basis, that the food system provides the total population access to a timely, reliable and nutritionally adequate supply of food."
Kennes 1990	"The absence of hunger and malnutrition."
UNICEF 1990	"The assurance of food to meet needs throughout every season of the year."
Mellor 1990	"The inability … to purchase sufficient quantities of food from existing suppliers."
Gillespie and Mason 1991	"The self-perceived ability of household members to provision themselves with adequate food through whatever means."
Von Braun 1991	"(Low) risk of on-going lack of access by people to the food they need to lead healthy lives."
Weber and Jayne 1991	"A situation in which an individual in a population possesses the resources to assure access to enough food for an active and healthy life."
Jonsson and Toole 1991	"Access to food, adequate in quantity and quality, to fulfil all nutritional requirements for all household members throughout the year."
ACC/SCN 1991.	"Access to the food needed for a healthy life for all its members and…not at undue risk of losing such access."

Alamgir and Arora 1991	"Enough food available to ensure a minimum necessary intake by all members."
Frankenberger and Goldstein 1991	"The viability of the household as a productive and a reproductive unit (not) threatened by food shortage."
FAO 1993	"Ensuring that all people at all times have both physical and economic access to the basic food they need."
FAO 1996	"Food security exists when all people, at all times, have physical and economic access to sufficient, safe and nutritious food to meet their dietary needs and food preferences for an active and healthy life."

not hold water if it is not linked to the important issues of access and entitlement suggested by Sen (1981).

The argument that arises from this shift to the household and individual is, what should the central focus be, the household or the individual? Some researchers have argued for the households as the unit of analysis (Eide 1990; Frankenberger and Goldstein 1991), while others (Gittinger *et al.* 1990; Reutlinger 1985) favour the individual, on the grounds that there is an issue of power and control in resource allocation. Within this argument, however, it would also be interesting to establish whether male or female individuals should be of central interest as drivers of food security.

In Uganda, for example, men are dominant in determining how resources are allocated and control the output, yet women have access to productive resources and spend more time on household farming activities, but have limited power and control over resources and output. This however, must await further analysis. Many definitions, including the commonly cited definitions by the World Bank (1986) and FAO (1996), begin with individual entitlement, implying the inevitable linkages between individuals and households and national and global economies.

From "food first" to "livelihood" perspectives

This shift reflects a change from the conventional view of food as a primary need, a lower order need as proposed by Abraham Maslow (1954). The shift focuses to building resilience into livelihoods to ensure that individuals and families have an economic base that allows them access to and consumption of quality food all year round.

In the shift from the food first to a livelihood perspective, food security is looked at in terms of future access to and availability of food. Sen argues in his entitlement approach that people sometimes choose to starve rather than sell their productive assets in order to build a livelihood for the future (Sen 1981:80). In the context of the 1984–85 famine in Sudan, in which over 100,000 people perished, De Waal (1991:68) also argues for livelihood security. He points out that in Darfur, people chose to go hungry in order to preserve seed for planting, cultivate their own fields or avoid selling an animal. He concluded that avoiding hunger is not a policy priority for rural people faced with famine: some people choose to go hungry today to have something to eat tomorrow. This is common practice, especially among women in Uganda, who will skip meals in times of scarcity to provide for others in the family.

This perspective has, however, not gone uncriticised. Chambers (1997) argues that people going hungry today to avoid going hungry later puts livelihoods at stake, as it focuses on objectives other than nutritional adequacy. Davies (1996) pinpoints the differences between the food first perspective and sustainable livelihood perspective (see Table 3). The notion of livelihood can be questioned further: for example, for how long should people go hungry in order to build resilient livelihoods tomorrow? This raises the question of time in analysing food security, a question that lies beyond the scope of this study.

Common forms of food insecurity

The World Bank (1986) has arrived at a conventional agreement about the distinction between chronic and transitory food insecurity. According to the International Fund for Agriculture Development (IFAD) (2014), chronic food insecurity is a food consumption trend that involves the inability to meet food requirements over a long period, while transitory food insecurity relates to

Table 3: Differences between narrow "food first" and wide "sustainable livelihood" approaches to household food security

Livelihood	"Food first"	"Sustainable livelihood"
Objective	Access to food	Secure and sustainable livelihood
Point of departure	Failure to subsist	Success in feeding, living
Priorities	Food at the top of the hierarchy of needs	Food one part of a jigsaw of livelihood needs
Time preferences	Food needs to be met before and in preference to all others	Food needs met to the extent possible given immediate and future livelihood needs
Entitlements	Narrow entitlement base (current and past consumption)	Broad entitlement base (includes future claims, access to common property resources, etc.)
Vulnerability	Lack or want of food	Defencelessness, insecurity, exposure to risk, shocks and stress
Security	Opposite of vulnerability is enough food, irrespective of the terms and conditions under which it is acquired	Opposite of vulnerability is security
Vulnerable groups	Based on social, medical criteria	Also based on economic, cultural criteria
Coping strategies	Designed to maximise immediate consumption	Designed to preserve livelihoods
Measuring and monitoring	Present and past consumption	Livelihood security and sustainability
Relationship to food security and environment	Degrade environment to meet immediate food needs	Preserve environment to secure the future

Source: Davies (1996)

shocks that briefly push the level of food consumption below requirements. A household can be said to be food secure only if it has protection against both kinds of insecurity. In my opinion, the question of going hungry today in order to avoid going hungry tomorrow can apply to transitory food insecurity and not chronic food insecurity. Choosing to go hungry in cases of chronic food insecurity weakens the ability of individuals and households to manage risks and vulnerability.

Analyses of transitory food insecurity look at intra- and inter-annual variations in household access to food, which according to CIDA (1989:21) can be categorised in two ways:
• Temporary food insecurity; occurring for a short time as a result of unforeseen and unpredictable causes; and
• Cyclical or seasonal food insecurity; with inadequate food access occurring at regular periods, which could result from a household's limited financial base or poor storage facilities.

As a result of these differences in food insecurity and the variations in the impacts and risks for different individuals and households, coping strategies also differ. Frankenberger and Goldstein (1990:22) argue that people have different patterns of responding/coping, depending on the nature of risk. They suggest that different household assets will play different roles in the process of coping. They thus conclude that small farming households face a fixed dilemma that involves a trade-off between immediate subsistence and long-term sustainability.

IFAD (2014), however, established that the household asset base will play an instrumental role in decisions to acquire and utilise food. A household with several assets can more effectively maintain its consumption levels by disposing of some of them. Its ability to do so increases according to the proportion of assets held in liquid form. Thus, the value and liquidity of assets are important determinants of a household's ability to cope with shocks.

Scholars such as Oshaug (1985) have thus identified three categories of household: first, *enduring households*, which maintain household food security on a continuous basis; second, *resilient households,* which suffer shocks, but recover quickly; and third, *fragile households*, which become increasingly insecure as a result of shocks. However, it can also be argued that hunger/food insecurity has a seasonal pattern: it follows the rhythms of harvest, seasonal availability and food price rises, as more and more partly self-provisioning people come on to the market (Kessy *et al.* 2013:99). When food is short, the number of meals eaten a day is reduced, as is the quality of the meals. This happens until days pass without anything to eat.

Kessy and others observed that small farmers either do well or badly out of the highs and lows of seasonal prices. For example, those that can store food till prices are high, do well, while those who have to buy when prices are high, do badly (Kessy *et al.* 2013). This implies that being able to produce food for the year and sell some, or to market cash crops, remains a key indicator of household well-being. In short, some households are able to acquire and utilise food, and others not, depending on their preparedness.

According to IFAD (2014), food insecurity in a household can be understood as a combination of two distinctive problems: a problem of *acquirement* and a problem of *utilisation*. Acquirement refers to the ability of a household and its members to acquire enough food through production, exchange or transfer. However, acquirement is only one aspect of food security or insecurity. A household with the capacity to *acquire* all the food it needs may not always have the ability to *utilise* that capacity to the fullest (IFAD 2014). For example, if a woman in Uganda, who is responsible for preparing and serving food to the family, finds herself burdened with responsibilities and unable to prepare and serve food in a manner that yields the best nutritional value, the family may experience transitory food insecurity. Yet in a country like Norway, where both husband and wife share the responsibility of preparing meals, this may not be a problem.

Alternatively, where the household lacks storage facilities to maintain adequate quantities of food in good condition outside the harvest season, there is the likelihood the household may become food insecure, even if it had solved the problem of acquirement by producing much. This study established that access to proper food storage facilities was still a big problem for smallholder farming households in Uganda. Therefore, a household can be said to be food secure only if it is secure in terms of both acquirement and utilisation of food.

Analysis of determinants of food security

IFAD (2014) suggests a fourfold conceptual framework for analysing food security. It looks at the ability to improve and maintain the "level of acquirement"; the ability to cope with "shocks to acquirement"; the ability to improve and maintain the "level of utilisation"; and the ability to cope with "shocks to utilisation".

The main determinants of the level of food requirement include endowment and entitlement mapping (Sen 1981; Osman 1995). On the other hand, the determinants of ability to cope with shocks to acquirement mainly relate to 1) reduction in fluctuations in household income, such as the degree of diversification of the household's livelihood strategy; and 2) reduction in fluctuation in consumption based on income fluctuations. This will require a household to establish the scope for consumption-smoothing: that is, the ability of a household to maintain the normal level of food consumption in the face of income shock (Hamound 2010; IFAD 2014).

Global and regional food security

FAO (2013:8) estimates that a total of 842 million people, or around one in eight people in the world, suffered from chronic hunger between 2011 and 2013, meaning they were regularly not getting enough food to conduct an active life. This figure shows a 3.08 per cent reduction in the level of hunger reported by FAO in 2011–12, when 868 million people were estimated to be undernourished.

It is important to note that there have been marked regional differences in this improvement in nutrition or food security. Sub-Saharan Africa remains the region with the highest prevalence of undernourishment, with modest progress in recent years. Western Asia shows no progress, while Southern Asia and Northern Africa show slow progress (see Figure 5). Many countries still struggle to meet the ambiguous food security goals of: 1) the World Food Summit (WFS) of 1996, whose target was halving the number of hungry people in the world (FAO 1996); and 2) the 2001 Millennium Development Goal (MDG) of halving the proportion of hungry people in the total population.

Although many countries still struggle to meet these targets, on 12 June 2013 FAO reported that 38 countries had already met internationally established targets in the fight against hunger, chalking up success ahead of established deadline of 2015 (FAO 2014). An additional 18 countries were congratulated on achieving both MDG 1 and the stricter WFS goal, having halved the absolute number of undernourished people between 1990–92 and 2010–2012. This is a positive indication that international food security targets are being achieved, and if other countries meet their commitments, it is possible that food security in Africa will improve steadily.

Despite the reduction in the world's hungry people and the successes of some countries in food security, the number of undernourished people remains high. According to the FAO chief: "Globally, hunger has declined over the past decade, but 870 million people are still undernourished, and millions of others suffer the consequences of vitamin and mineral deficiencies, including child stunting" (FAO 2014a: para 4).

Substantial reductions in both the estimated numbers of the undernourished and the prevalence of undernourishment have occurred in most countries of Eastern and South Eastern Asia, as well as in Latin America. However, the situation is still worse in South Asia, closely followed by sub-Saharan Africa, as shown in Table 4 below.

Justification for regional differences in undernourishment

FAO observes that several factors account for the regional differences in hunger reduction, including differences in economic conditions, infrastructure, the organisation of food production, the presence of social provisions and political and

Table 4: Undernourishment around the world, 1990–92 to 2011–13

| | Numbers of undernourished (millions and prevalence (%) of undernourishment) | | | | |
	1990–92	2000–02	2005–07	2008–10	2011–13*
WORLD	1 015.3	957.3	906.6	878.2	842.3
	18.9%	*15.5%*	*13.8%*	*12.9%*	*12.0%*
DEVELOPED REGIONS	19.8	18.4	13.6	15.2	15.7
	<5%	*<5%*	*<5%*	*<5%*	*<5%*
DEVELOPMENT REGIONS	995.5	938.9	892.9	863.0	826.6
	23.6%	*18.8%*	*16.7 %*	*15.5%*	*14.3%*
Africa	177.6	214.3	217.6	266.0	226.4
	27.3%	*25.9%*	*23.4%*	*22.7%*	*21.2%*
Northern Africa	4.6	4.9	4.8	4.4	3.7
	<5%	*<5%*	*<5%*	*<5%*	*<5%*
Sub-Saharan Africa	173.1	209.5	212.8	221.6	222.7
	32.7%	*30.6%*	*27.5%*	*26.6%*	*24.8%*
Asia	751.3	662.3	619.6	585.5	552.0
	24.1%	*18.3%*	*16.1%*	*14.7%*	*13.5%*
Caucasus and Central Asia	9.7	11.6	7.3	7.0	5.5
	14.4%	*16.2%*	*9.8%*	*9.2%*	*7.0%*
Eastern Asia	278.7	193.5	184.8	169.1	166.6
	22.2%	*14.0%*	*13.0%*	*11.7%*	*11.4%*
South-Eastern Asia	140.3	113.6	94.2	80.5	64.5
	31.1%	*21.5%*	*16.8%*	*13.8%*	*10.7%*
Southern Asia	314.3	330.2	316.6	309.9	294.7
	25.7%	*22.2%*	*19.7%*	*18.5%*	*16.8%*
Western Asia	8.4	13.5	16.8	19.1	20.6
	6.6%	*8.3%*	*9.2%*	*9.7%*	*9.8%*
Latin America and the Caribbean	65.7	61.0	54.6	50.3	47.0
	14.7%	*11.7%*	*9.8%*	*8.7%*	*7.9%*
Caribbean	8.3	7.2	7.5	6.8	7.2
	27.6%	*21.3%*	*21.0%*	*18.8%*	*19.3%*
Latin America	57.4	53.8	47.2	43.5	39.8
	13.8%	*11.0%*	*9.0%*	*8.0%*	*7.1%*
Oceania	0.8	1.2	1.1	1.1	1.2
	13.5%	*16.0%*	*12.8%*	*11.8%*	*12.1%*

Source: FAO, IFAD and WFP (2013:8)

Note: * Projections

institutional stability. The slow progress in reducing hunger in sub-Saharan Africa, to which Uganda belongs, can be attributed to often miserably inadequate infrastructure, a problem that plagues vast areas of rural Africa (FAO 2013). Improved communications and broader access to information technology may, to some extent, account for increased hunger reduction rates in the East Asia and some parts of North Africa.

Uganda's food security is still not the best compared to the rest of the world and sub-Saharan Africa. Near 1.4 million people, approximately 3.9 per cent of the total population, are food insecure (FAO, IFAD and WFP 2013), and the number of Ugandans experiencing food insecurity in terms of caloric intake is alarming (Ssewanyana and Kasirye 2010). The country has registered increased undernourishment since the early 2000s and the hope of achieving the MDG hunger target by 2015 is minimal.

In the past few years, there have been several attempts to study food insecurity in Uganda in terms of its nature, extent and impacts. The studies include those by the World Food Programme (2013 and 2009), Simler (2010), Ssewanyana and Kasirye (2010 and 2003), Pouw (2009), Benson *et al.* (2008), Alderman (2007), Ssewanyana *et al.* (2006) and Banhingwa (1999). All of them reveal an unfavourable food security situation, coupled with increasing food prices.

Uganda's growth in food production has not kept pace with population growth, at an annual rate of 3.2 per cent. Statistics show that food production per capita has been declining since 2002–04. Additionally, dietary energy supplies, including the energy supplied by imported food, has kept declining since 2003–05, although it remains, on average, adequate to meeting the energy requirements of the population.

What has the government done?

Uganda has had low adoption and use of modern agricultural technology, which has contributed to low productivity growth. The government of Uganda is, however, trying to tackle this through the implementation of the National Agricultural Advisory Services (NAADS), established in 2001 as a public-private approach to extension service delivery. One of NAADS's key objectives is promoting food security, nutrition and household incomes through increased production and market-oriented farming (MoFPED 2012:19). The second phase of NAADS, NAADS II, launched in December 2011, is being implemented under the Agriculture Technology and Agribusiness Advisory Services (ATAAS) umbrella, and seeks to generate and disseminate technology under the National Agriculture Research Organisation, NARO (MoFPED 2012:18).

Although NAADS has been effective in reducing food insecurity through the distribution of various inputs and technology to farmers across the country, it has not been effective in reaching resource-poor farmers, the majority of whom are engaged in subsistence agriculture. In the eastern region, as in other regions, farmers with between two and five acres benefited more by receiving NAADS technology than those with less than two acres (MoFPED 2012:57).

It is also evident that about 41 per cent of the food-secure farmers served by the NAADS programme sell 70–100 per cent of the output produced using the NAADS input. This scenario illustrates the ineffectiveness of the programme in ensuring food security (MoFPED 2012:63).

The government of Uganda has also made more commitments to attaining food security. For example, Uganda ratified Article 25 (1) of the Universal Declaration of Human Rights and Article 11 (1) of the International Convention on Economic, Social and Cultural Rights in 1987 (MoFPED 2012:10). Both of these provide for the right of all to adequate standards of living, including adequate food. The importance of food and nutrition is recognised in *Objectives XIV*, and *XXII* of the 1995 Constitution of the Republic of Uganda (Republic of Uganda 1995).

In this respect, the Uganda Food and Nutrition Policy (UFNP), which explicitly recognises the right to food, was developed and adopted in July 2003 by a multisector Food and Nutrition Council (UFNC), under the leadership of the Ministry of Agriculture, Animal Industry, and Fisheries (MAAIF) and the Ministry of Health (MoH), the line ministries responsible for its implementation and coordination with other stakeholders (MAAIF and MoH 2003; MoFPED 2012). UFNP was framed within the overall national development policy objective of *poverty eradication*, as described in the Poverty Eradication Action Plan (PEAP), which serves as the framework for all national development policies in Uganda (MAAIF and MoH 2003; MoFPED 2005).

However, there has been no remarkable improvement in the nutrition levels of many Ugandans since the adoption of the UFNP. The prevalence of malnutrition in Uganda was estimated at 26.3 per cent in 2000–02, and stood at 30.1 per cent in 2011–13 (FAO, IFAD and WFP 2013:39). This implies that the adoption and implementation of UFNP have not been effective in addressing malnutrition and food insecurity.

In March 2010, the government also signed the comprehensive Africa Agriculture Development Programme (CAADP) compact, committing itself to allocating 10 per cent of the national budget to agriculture and pursuing policies to ensure a growth rate of 6 per cent per annum for the sector (MoFPED 2012:10). However, in the same financial year, government spending on agriculture amounted to only 5 per cent of the total budget, down from 7.6 per cent in the previous year (FAO, IFAD and WFP 2013:39). Budget allocations to agriculture have kept declining and have never met the African Union Maputo Declaration of allocating 10 per cent to agriculture. This shows a lack of government commitment in fighting food insecurity and poverty among its people.

The continued low investment in agriculture and agricultural technology has steadily affected the share of agriculture in total GDP. It has declined from, for example, 18.3 per cent in financial year 2005-06 to 13.9 per cent in financial

Table 5: Sectoral real GDP growth rates (per cent)

Sector	2005/06	2006/07	2007/08	2008/09	2009/10	2010/11
Agriculture, forestry and fishing	0.5	0.1	1.3	2.9	2.4	0.9
Industry	14.7	9.6	8.8	5.8	6.5	7.5
Services	12.2	8.0	9.7	8.8	7.4	8.0
GDP at Market Price	10.8	8.4	8.7	7.3	5.5	6.3

Source: UBOS data, MoFPED (2011)

year 2010–11 (MoFPED 2013:11). The growth rate of the agriculture sector in Uganda has been erratic and on a declining trend from 2.9 per cent in 2008–09 to 0.9 per cent in 2010–11, amidst high population growth (3.2 per cent) (see Table 5).

The fluctuating and deteriorating growth of the agricultural sector in Uganda shows the sector's low level of sustainability. People produce only what can sustain them for a short time, but thereafter go hungry again. This pattern cannot aid the achievement of sustainable development. Efforts need to be made to enable the growth of the agricultural sector beyond mere sustainability towards sustainable development. Per capita food production in Uganda is much more variable than the average for sub-Saharan Africa, largely as a result of lack of irrigation.

Uganda's agriculture is heavily rain-fed, with less than one per cent of land under irrigation. Rain-fed agriculture refers to farming practices that rely on rainfall for water (Rosegrant *et al.* 2002). This leaves many farmers vulnerable to climate variability. Given increased rainfall variability and more frequent extreme events over the last decade, some parts of Uganda have experienced severe food shortages. In the northeast, for example, inhabited by the Karamojong, consecutive years of poor weather and below normal rainfall since 2012 have had severe impacts on food security in the area.

In June 2013, food security partners led by MAAIF undertook an Integrated Food Security Phase Classification (IPC) analysis and released their report in November. It revealed that up to 1,194,423 people in the semi-arid Karamoja region faced stressed levels of food insecurity. Although the Karamoja had enough food stocks at community level, they faced challenges of inadequate food intake. Levels of malnutrition were high due to poor dietary diversity, childcare and feeding practices. It was predicted that these people would in the near future have to purchase food from the market with income obtained from the sale of livestock, firewood and charcoal (IPC 2013).

In almost the same period, the Famine Early Warning Systems Network (FEWSNET) reported that people adopted negative coping strategies that involved consuming stocks sparingly by reducing the number of meals and skipping meals to get through the lean season, but that food was, nonetheless, exhausted by March 2014. The demand for food from markets increased and other coping strategies are being used, such as food loans, sharing assistance and food, begging, and abnormally high reliance on firewood and charcoal sales (FEWS-NET 2014). This has left many people highly indebted, and heavy firewood and charcoal sales have negative implications for environmental conservation and sustainability.

FAO, IFAD and WFP (2013) recommend that for Uganda to realise its agricultural potential, the government must provide public goods such as extension services and irrigation, transport and communication infrastructure to permit smallholder farmers, who account for over 95 per cent of all farms, to increase their productivity.

Food security in Uganda, particularly the eastern region

Uganda is endowed with large fresh-water resources, favourable soil conditions and a climate with great agricultural potential. It is estimated that about 81 per cent of all households (42 million) are engaged in agriculture. However, agricultural involvement by households varies from region to region. Nighty-one per cent of households in Uganda's eastern region are actively engaged in agriculture (WFP 2013:6).

According to FAO statistics (WFP 2013:6), Uganda produces enough plantain and cassava, the two most important staples, to feed its people. Surplus maize and beans are also produced, which enables exports to nearby Kenya and South Sudan (see Table 6).

Matooke (banana) is mainly produced in central and western Uganda, while the east produces more maize, and the north produces fewer food crops (maize, bananas, beans and sweet potatoes) than other regions, given its aridity. Generally, matooke accounts for the largest share of expenditure on food as well as the largest median quantity consumed (Uganda Bureau of Statistics, 2006; Ssewanyana and Kasirye 2010:11). Other main foods include sweet potatoes, cassava, rice, maize grain and flour, bread, fresh and dry beans, millet and sorghum.

Mixed agriculture is common in Uganda and livestock production is an in-

Table 6: Production of major crops (1,000 metric tons) by region

Crop	Central	Eastern	Northern	Western
Maize	712	948	376	548
Beans	263	180	95	314
Cassava	471	435	415	116
Bananas for food	4,296	239	35	3,430
Sweet potatoes	423	458	278	313

Source: World Food Programme (2013)

Table 7: Distribution of households that took one meal a day (per cent)

Residence	2002/03	2005/06	2009/10
Urban/Rural			
Rural	6.0	9.0	10.1
Urban	8.1	6.3	5.9
Region			
Kampala	5.3	6.4	6.9
Central	3.7	9.6	7.3
Eastern	3.0	4.8	7.3
Northern	25.1	18.1	20.1
Western	4.5	3.8	5.8
Uganda	7.7	8.5	9.3

Source: Uganda Bureau of Statistics 2010

tegral part of the agriculture sector, contributing up to 5.2 per cent of the country's GDP (WFP 2013:7). In eastern Uganda, up to 83 per cent of households engage in livestock production alongside crop production, followed by 82 per cent in the northern region

Climate variability resulting from prolonged dry spells, however, affects livestock and food production in parts of Uganda, particularly the east and northeast, affecting access to pasture and water, as well as reducing the number of meals consumed per day (WFP 2013; FEWSNET 2010).

Traditionally, homes in Uganda take three meals a day. A meal according to the Uganda National Household Survey (UNHS) (Uganda National Household Survey 2006) was considered to be a substantial amount of food eaten at one time. The 2009/10 UNHS findings (Uganda Bureau of Statistics 2010), however, indicate substantial reduction in the number of daily meals, with many families taking one meal a day. This was pronounced in rural areas most affected by food insecurity, especially northern and northeastern Uganda (see Table 7). This implies significant variation in food security across the country that partly results from the sources of food and climate conditions.

Sources of food and food security in Uganda

Although most households are involved in farming, Ugandans are fairly market dependent, with markets being the main source of food calories for about 50 per cent of households. This makes many households vulnerable to food insecurity when food prices rise sharply (WFP 2013). The seasonality of the cropping calendar, accentuated by climate variability, has been one of the major causes of high food prices in Uganda.

Mount Elgon ecological zone has a subtropical "bimodal" climate with two rainy seasons (March-May and July-September), followed by dry seasons. Thus, there are two crop-growing seasons, with the first harvest usually occurring between June and August and second from November to January. Of late, however, climatic patterns have been unreliable, with rains coming later than expected, which affects food production.

Because of lack of proper storage facilities, limited access to credit and sources of income, smallholder farmers in Uganda are often compelled to sell their food surpluses on the market immediately after harvest (WFP 2013). Sometimes, they sell directly to established markets, at others through intermediaries (middlemen), who pay small prices. As a result, the food market chain in Uganda is long.

The WFP plays a major role in the grain market of Uganda as a wholesaler and buyer. Although the grain market looks to be vibrant, men are the key players and women, who make a crucial contribution as farmers, workers and entrepreneurs, are disadvantaged due to gender gaps. The gender-related issues in Uganda's food security are discussed below.

The sources of food consumed in Uganda are usually categorised by type of food. For example, most consumers of cereals, roots and tubers grow them themselves, while meat, dairy products and vegetables are mainly purchased from the market (Ssewanyana and Kasirye 2010:14). Access to, source and category of food also vary by region. Most unprocessed staples are purchased. For example, in the eastern region, the majority of households produce their own root and tuber crops and cereals (maize, sorghum and millet), which are widely consumed. Other foods like fruits, vegetables, meat and dairy products, except chicken, are mainly purchased from the market (Ssewanyana and Kasirye 2010).

Gender and food security in Uganda

In 2010, Uganda had 6.2 million households, 30.1 per cent of them female headed (UBOS 2010). Despite this significant percentage, women still face gender-specific constraints that impair their productivity as well as reduce their contribution to agricultural production, economic growth and family and community well-being.

For example, women face gender constraints in accessing productive resources. They control less land than men, and what they do occupy is often of poor quality, yet they have less credit to obtain modern fertilisers, pesticides and improved seeds than their male counterparts. Land tenure is also insecure, due to custom and economic conditions. Control of such credit as women do obtain from their produce is in the hands of men, especially in the case of married women. Yet it is believed that if women had equal access to good quality resources, their farm output would equal that of men, and the level of undernourishment in Uganda and other developing countries would decline (FAO 2011b).

Box 2: Summary of food security in Uganda

- Nationally, almost half (48 per cent) of Ugandans were food energy deficient between September 2009 and August 2010.

- Nearly 5 per cent of Ugandans had poor food consumption, which reflects an extremely unbalanced diet that is devoid of protein and chiefly comprised of starchy maize or matooke (plantain) flavoured with vegetables. Seventeen per cent had borderline food consumption, which means they consume a slightly more varied diet with more pulses, vegetables and sugars, but still barely any animal proteins, milk or fruit.

- A third of Ugandan children were stunted, 14 per cent severely so, and the rate was "serious" in western (42 per cent) and eastern (36 per cent) Uganda. Rural Ugandans were also more likely to be stunted than urban (37 per cent v. 14 per cent).

- Food insecurity and malnutrition are strongly associated with monetary poverty (here measured by the expenditure quintile). Despite Uganda's progress in reducing poverty, the absolute number of poor people has increased due to population growth, and poverty remains firmly entrenched in rural areas. About 30 per cent of rural people still live below the national rural poverty line.

- The poorest households in rural Uganda were the most dependent on purchasing food, making them highly vulnerable to food price rises.

Source: Adapted from World Food Programme (2013:1)

Summary of food security in Uganda

WFP (2013), during the Comprehensive Food Security and Vulnerability Analysis (CFSVA) in Uganda, came up with ten key findings, five of these are highlighted in this study, which give a clear picture of Uganda's food security status (see Box 2).

Impact of climate change on food and livelihood security

The increased incidence of flooding, extreme drought and rising sea levels have been proven to have "immediate impacts on food production, food distribution infrastructure, the incidence of food emergencies, livelihoods, assets and opportunities and human health in both rural and urban areas" (FAO 2007:7). These impacts have a direct influence on the socioeconomic development of a society. The impact of climate change on food security not only takes the form of drought or floods, but also of water security and the incidence and spread of pests and diseases that affect humans, plants and animals (Lesley 2008:3).

Evidence from the IPCC (2000–07) indicates that countries in temperate regions are likely to enjoy some economic advantage from climate change, because additional warming will benefit their agriculture sector. However, countries lying in tropical and subtropical regions are predicted to be more vulnerable to warming, which will affect the water balance and harm the agriculture sector. The IPCC predicts that the worst affected region will be Africa, where there is already evidence of the severe effects of climate change. Many countries in Africa, including Uganda, lack updated information systems and have poor technology, while their economies largely depend on agriculture.

Farmers in Africa already face dwindling water supplies and water variability due to unstable rainfall. This has led to millions of people being affected by drought. The impact is predicted to be more acute as a result of climate change (Dinar *et al.* 2008). Agriculture, just like fisheries and forestry, is very sensitive to climate and increased warming caused by climate change will likely affect production. FAO (2008:11) identifies two main food security implications of changes in agricultural production:

1. Impacts on the production of food on the global and local levels: low income countries that have limited financial capacity to trade and depend highly on production of their own food will suffer more compared to developed regions that can easily offset declines in local supply through imports.
2. Impacts on all forms of agricultural production will affect livelihoods and access to food. This implies that producer groups like rural and smallholder farmers in developing countries, who are less able to deal with climate change, risk having their safety and welfare compromised.

Adaptation by smallholder farmers to climate change in Uganda
Smallholder farmers in Uganda are using several strategies to adapt to climate change.

Encroachment on swamps
In cases of the increased incidence of drought and moisture stress, farmers are encroaching on swamp areas to grow crops suitable for swamp conditions, such as potatoes and rice (Bagamba *et al.* 2012). This implies shifting labour from other crops. However, Bagamba *et al.* found that encroachment on swamps did not result in economic gains. First, the acreage under swamp cultivation was too small to have a significant impact, and second, shifting resources from a higher value crop (bananas) to sweet potatoes was not economical. Instead, swamp encroachment negatively impacted the wetland resource and its ecosystem.

Crop-livestock integration
Farmers are minimising the risk of crop loss by diversifying. They grow many different crops and engage in non-farming activities like fishing, hunting and gathering wild food plants (Bagamba *et al.* 2012). This also includes changes in the cropping times, based on the availability of rains; changes in production techniques; as well as growing drought-resistant varietals. These findings align with what Diner *et al.* (2008) found out about adaptation strategies by smallholder farmers elsewhere.

The above strategies suggest that some adaptation by farmers takes place autonomously and may not need government or policy intervention. However, government could have an important role in promoting, for instance, infrastructure improvement, providing weather information and stabilising local market conditions.

Other non-governmental organisations and projects like food banks also have roles to play, especially as many poor people turn to them for services and inputs. It is useful to identify what role such agencies play or could play in helping smallholder farmers achieve food security, build sustainable livelihoods and also adapt to climate change. The concept of the food bank has scarcely been researched by scholars.

Smallholder farmers, however, face numerous limitations in effective adaptation to climate change. They experience poverty and lack of credit; have limited market access and problems with transport; have difficulty in obtaining drugs for their cattle and seeds for drought resistant plants; are reluctant to grow crops they cannot themselves consume, due to poor market access; lack information on appropriate and efficient adaptation and the weather, particularly the timing of the rains; and perceive many adaptation strategies as expensive (Dinar *et al.* 2008).

Tarasuk (2005:303) argues that food insecurity can force people to adopt food consumption patterns and employ a variety of strategies to acquire food or money that fall outside social norms. Food insecure households make daily decisions about how much of the household expenditure can go towards food, what foods should be purchased, how many meals to have a day and so on.

While food insecure households in developing countries may opt to skip meals to save food for the next day, and resort to food loans, sharing and begging, etc. (FEWSNET 2014), food insecure households in developed countries seek food aid from food banks and other emergency food providers (EFPs) (Moldofsky 2008; Berner and O'Brien 2004). Some of these coping strategies may be conscious, while others may be ad hoc, in response to changing circumstances (Barnett 2001). This study, however, focuses on the food bank, which has spread from developed to developing countries.

What are food banks?

Whereas some food banks provide food directly to hungry people, others in the Food Bank Movement see a food bank as "… a centralized warehouse or clearing house registered as a non-profit organization for collecting, storing and distributing food (donated/shared), free of charge, to front line agencies which provide supplemental food and meals to the hungry" (Riches 1986:16). In this definition, emphasis is on the notion of surplus food in the food production and retail system. It also says much about the relationship between surplus food and the hungry. This signifies that food banks intend to make food that would otherwise be dumped/wasted available to organisations that can put it to good use. Therefore, the term "food bank" includes food depots, food pantries and other community-based food distribution sites (Starkey *et al.* 1998:1144).

The Food Bank of Singapore (2013), on the other hand, looks at itself as a place where companies or people can come to deposit their unused or unwanted foods, which will then be allocated to the needy via Voluntary Welfare Organisations (VWOs), charities, soup kitchens, etc. This definition also stresses the notion of surplus and unwanted food and the needy.

However, while some argue that it is difficult to distinguish between surplus food and waste and the needs of the hungry people, others feel that "calling donated food" "waste" has a harsher connotation (Riches 1986:16). Beyond that, surplus would suggest to the public that this is food that cannot be used by the food industry, whereas waste leaves the impression that the private sector is inefficient.

Generally, food banks can be understood as community-centred warehouses that solicit, store and distribute food from local producers, retail food sources, federal community distribution programmes and the food industry (Nicholas-Casebort and Morris 2001).

Who are food bank users in developed countries?

Most studies reveal that food bank users are diverse, with diverse characteristics, and problems. However, the common factor is that they are usually food insecure people who also often face financial insecurity. Studies in New Zealand, for example, reveal that frequent bank users were recipients of state benefits (NZC-CSS 2005; Thériault and Yadlowski 2000; Mackay 1995). These findings are in line with Starkey *et al.* (1998), who found that 83.5 per cent of food bank users in Montreal, Canada, were also in receipt of social assistance.

Another main category of food bank user is low income earners and the unemployed. The former often had challenges in meeting their food requirements as well as livelihood security, because of lack of job security as well as increasing poverty among their households. They thus resorted to food banks for food assistance (Stephen *et al.* 2000).

Riches (2002) and Uttley (1997) found that single mothers were also frequent food bank users, partly due to lack of stable incomes, although other factors may have contributed.

Other studies point to immigration as contributing to the increased number of food bank users. Starkey *et al.* (1999) and Daily Bread Food Bank (2005) in Montreal and Toronto, for example, indicate that nearly 50 per cent of food bank clients are immigrants, mainly from Eastern Europe, South America and the Caribbean. The author of this study can confirm from experience that frequent users of the Salvation Army food pantry in Kristiansand, Norway are mainly immigrants from Nigeria, Liberia, Somalia, Poland, Eritrea and other lands. He therefore agrees with the above findings by Starkey *et al.* and the Daily Bread Food Bank.

Who are food bank providers?

Food banks are mainly run by NGOs and/or VWOs, which are often founded by or affiliated with religious organisations (Fitzgerald and Cameron 1989:23; Crack, *et al.* 2007; Daly 1996). Other food banks, such as the Hunger Project food bank, have been founded by non-profit charitable organisations (Global Hunger Project 2012). Traditionally, food banks serve two main purposes: assisting low-income consumers and distributing surplus food. Although many of these organisations offer food assistance, they have diverse objectives and food banks are often just part of their programmes.

McPherson (2006) observed that many food banks are used by operators to encourage users to access other services they offer, for example advocacy and budgeting services or general life-skills courses. She notes that after a certain number of visits to the food bank, it may be a requirement that the client seeks budgeting advice. In some cases, the food bank services are used as a means to generate demand for other services that are supported through a contractual agreement with the state.

The present author has visited the Salvation Army pantry in Kristiansand on many occasions since commencing this study in 2013. He observed that the providers used the food pantry to encourage people to listen to the word of God. Usually while people are eating and taking coffee in the service hall, a team of church members will minister to them with religious songs and a brief sermon. Users are also encouraged to attend church services with them if they have no church of their own. It can therefore be argued that whereas many food banks offered food aid to the hungry, many did not have the objective of ending hunger and promoting food security, but rather of providing relief and emergency food supplies to the hungry.

Can conventional food banks reduce food insecurity?

Riches (1986:4) argues that the rise of food banks provides concrete evidence of the collapse of social assistance and unemployment insurance. This implies the breakdown of the safety net at the very point in an economy where it should be providing strongest support to the most vulnerable.

According to Husbands (1999), food banks and other community assistance programmes should only be relied on as emergency measures, rather than being institutionalised as permanent mechanisms for food access. Husbands argued that traditional food banks are geared to emergency assistance (hunger alleviation) but do not address hunger as a structural phenomenon. They solicit food from the public and corporations in addition to purchasing it from producers. Because of different understandings of the operation of food banks, different models of food banks exist. However, the simple model employed by traditional food banks is the "hunger-alleviation model" (Husbands 1999:107).

Food companies, restaurants and individuals donate food to food banks. However, charities and churches are the major funders of food banks in the US (Hayes and Laundan 2009:428). Due to their dependency on public and corporate goodwill, many food banks are unwilling to pursue social change in any determined way (Husbands 1999). They often employ voluntary labour and very-low key fundraising, and usually have insufficient human and financial resources to undertake the research, advocacy and community mobilisation needed to address hunger systematically. While anxiety about the adequacy and appropriateness of this response to income-related food problems proliferates, food banks are now regarded as a "necessary community resource," especially in the US, for emergency food (Starkey *et al.* 1998:1148), and for farming inputs in Uganda.

Starkey *et al.* (1998) found that most food bank users in Canada were not those thought to be the most vulnerable in terms of nutritional status (the very young, elderly and those with chronic health conditions). They were rather healthy, single individuals, although also mainly the non-working poor. It has been observed that the assistance provided by food banks largely depends on the quality and quantity of donations from the public and from producers, processors and retailers (Teron and Tarasuk 1999:382). This affects the amount of food donated and its nutritional value. The amount of food supplied is usually not sufficient to meet the caloric requirements of food bank users.

Paradigm shift in the role of food banks in food security

Because of the recent rise in obesity and diet-related diseases among food insecure people, food bank personnel have been persuaded to actively promote more nutritious products (Handforth *et al.* 2013). Some food banks offer nutritional education to users (Food Bank of Delaware 2011). Others have engaged researchers to develop a nutrition-profiling system to measure the food distributed in terms of MyPyramid day (Anderson 1990).

Other measures have been developed to establish the exact cause of food insecurity, for example the Core Food Security Measure (CFSM). The CFSM was developed using a probabilistic log-linear measurement model, the Rasch model (Hamilton *et al.* 1997), which was developed by George Rasch (Rasch 1966). This measurement scale is used to assess the extent and severity of household food insecurity over the previous 12 months due to insufficient money for food (Derrickson, Fisher and Anderson 2000). This paradigm not only has the potential to address food insecurity and malnutrition in vulnerable populations, but also represents a new opportunity for anti-hunger and nutrition advocates to work together towards a common goal (Handforth *et al.* 2013).

Earlier on, in 1999, Husbands suggested that for food banks to provide the desired local solutions to food insecurity and build livelihood security, they needed to transform themselves into anti-hunger organisations. The expanded role of such organisations would include significantly reducing the incidence of hunger and eliminating the need for food banks as welfare or emergency agencies. This new role would require anti-hunger organisations to become involved in research, public education, public policy advocacy, one-on-one advocacy and community mobilisation. This does not, however, rule out the invaluable role played by food banks in providing emergency assistance programmes and services to meet people's basic needs (Husbands 1999:108).

Alternative approaches to the food bank

Elsewhere in the world, the concept of food bank has been adopted and implemented differently with the aim of addressing food insecurity and reducing the number of hungry people. In Cambodia, Laos, Thailand and Bangladesh, for example, the concept has been adapted in the form of "Community-managed Rice Banks" (Datta 2007; UNBconnect 2014), while in India, Cameroon and Burkina Faso, food banks take the form of "Community Grain Banks" (Inter Pares 2004; Edakkadi 2013; RELUFA 2008; Eugene 2013; Yamengo 2013). These institutions are reported to have had a great impact on the food security of these countries.

What are community-managed rice banks?

Generally, a rice bank is a simple wooden structure for housing rice. In many cases, the initial supply of rice in these banks comes from the community's surplus (from a collective rice field maintained for that purpose), or from external agencies, including local government (Datta 2007). Villagers borrow rice from the rice bank and repay the same amount when their next crop is harvested.

CARITAS (2004) and UNBconnect (2014) reported that rice banks significantly contributed to food security for the poor in rural areas. In Bangladesh, for example, UNBconnect reported that the Chittagong Hill Tracts (CHT) had always been the most food insecure region. However, by May 2014, a total of 1,708 community rice banks had been established in the hilly districts of Rangamati, Bandarban and Khagrachari under the community empowerment and economic development project financed by UNDP's Chittagong Hill Tracts Development Facility (CHTDF). It was reported that these areas had become food secure, with the rice banks allowing them to survive the "lean" period (UNBconnect 2014).

Management of community rice bank

Community rice banks are established and managed by community members themselves. They supply the materials for the construction and renovation of the rice warehouses, and only receive external assistance in the form of technical inputs and other material support (Datta 2007). Community members are helped to establish a three-person management committee comprising the chief (usually the village chief), clerk (accounting officer) and rice bank keeper.

The committee oversees loan disbursement, leads the village in fixing the annual interest rate on rice returned to the bank, and reports on activities, including transaction details. It is trained in the basic principles, objectives and management of rice banks, and provided with supplementary documents like stock accounts, loan applications and contracts, stock records, membership applications, and the format for monthly, quarterly and annual reports. Villagers are also sensitised to the benefits of managing rice banks in a sustainable manner and their role in achieving this sustainability (Datta 2007).

Inter Pares (2004) reported that rice banks were judged to be working well and ensured that villagers' families in the areas in which they were established did not go hungry. It was also reported that rice bank users were well motivated and that rice banks were sustainable.

What are community grain banks?

The community grain bank is a community-managed food security system, where the community is trained to set up a bank of food grains from which they can borrow during times of need and repay in kind with minimal interest fixed by the community (Swaminathan n.d.). Unlike the specialised community rice banks, grain banks are designed to store different grains at the same time, such as ragi, paddy/rice, maize and millet (Reddy and Adolph 2002; Carter 2001).

In grain-producing countries, grain prices are usually very low after the harvest, when most farmers have plenty of grain. Later in the year, prices rise sharply. When people's own grain supplies start to run out and the need to buy grain arises, grain prices are often high. To prevent food insecurity, grain banks buy grain when prices are low around harvest time and sell it at fair prices when it is in short supply. Grain banks have been proven to protect farmers/poor people during times of drought against exploitation by traders, who take advantage of difficult situations to sell grain to people at very high prices (Carter 2001).

This community-based food security strategy is considered very effective and has been replicated in many other districts of India. For example the Academy of Development Science (ADS) had by 2002 established 132 grain banks in 120 villages of Raigad and Thare districts, Maharashtra state (d'Silva 2014). Community grain funds/banks have not only supported women in achieving strong biodiversity on their farms, but also reestablished women's control and leadership over community food production knowledge (d'Silva 2014).

Learning from conventional food banks

Although many conventional food banks in developed countries may not be able to adapt alternative approaches such as "community-managed rice banks" and "community grain banks" because of contextual and other differences, some food banks have tried to shift from the traditional role of food collection and emergency food supplies to initiating community gardens, income-generating projects, skills training, linking farmers to markets as well as providing farmers with improved seeds. These roles are important in ensuring sustainable access to, and availability and utilisation of food, as well as in improving livelihood security. A sampling of food banks and their operations is presented to show how they could serve as practical examples in Uganda.

Food Bank of Delaware

On top of providing a children's nutrition programme, mobile pantry, school pantry and culinary school – all typical of traditional food banks – the food bank of Delaware runs a supplemental nutrition assistance programme (SNAP) in collaboration with government (Food Bank of Delaware 2011). It also runs a community-supported agriculture programme (CSA). CSA is a generic term coined in the US in 1985 to promote fresh, locally grown food and foster social and ecological responsibility. It can be defined as a partnership between farmers and consumers, where the responsibilities and rewards of farming are shared (Soil Association 1997:5)

CSA has become a popular way to buy local seasonal food. This not only allows money to remain in circulation within the local community, leading to local development, it also checks the buying of foods that are imported or transported from distant areas, with effects for global warming and the environment (NRDC 2010).

Furthermore, by encouraging CSA, the Delaware food bank helps farmers to avoid exploitation by moneylenders. In this case, CSA members become the shareholders of a local produce farm and the money they pay to farmers helps them buy seeds, pesticides and fertilisers and prepare the land, thereby avoiding loans (Food Bank of Delaware 2011; Soil Association 1997:5).

Community food banks in Uganda can emulate this and by initiating and promoting various models of CSA can support smallholder crop farmers in achieving food security and sustainable livelihoods. Models include the subscription or farmer-driven model, the shareholder or consumer-driven model, farmer cooperatives and farmer-consumer-cooperatives (Soil Association 1997:6).

CSAs have a direct effect on the accessibility and availability of local, nutritious and fresh foods for consumption by the community. One of the challenges facing Uganda's smallholder crop farmers is limited access to financial resources and markets. Implementing the CSA strategy would provide local solutions to this challenge.

Community Food Bank of Southern Arizona

This food bank offers several programmes and services. However, two were found interesting for the purposes of this study: the farm-to-child programme; and the community food consignment programme (Community Food Bank 2009). Through the former, the community food bank of Southern Arizona uses experts to help children, parents and teachers learn how to garden through technical assistance and workshops. It is noted that children involved in gardening show an increased preference for fruits and vegetables (Heim *et al.* 2009:1220), thereby improving their nutrition intake.

The community food consignment programme, on the other hand, promotes farmer productivity by offering gardeners and small farmers an opportunity to sell produce and other goods at farmers' markets. This helps to address one of the common problems facing small farmers – access to markets. It is an attractive programme, which encourages backyard gardening and chicken rearing not only for income, but also for improving family nutrition (Community Food Bank 2009).

Both these programmes could offer practical solutions to the food challenges in Uganda. Universal primary and secondary education institutions in Uganda, for example, struggle to feed pupils and students. Food banks in Uganda may need to team up with schools to develop school gardens as a productive and sustainable source of nutritious meals to students. Also, community gardening could be a useful way by which food banks in Uganda could help smallholder farmers, who rely on family labour, to expand their production. This could be achieved through joint farming and utilising the services of volunteers from agriculture universities/institutions in Uganda and from among food bank volunteers.

South Africa

Agencies that work as links between the food bank and local people are encouraged by the food bank to establish income generation and agriculture projects as well as skills training. These projects are monitored by the food bank of South Africa's community field workers. In 2013, of all the agencies that worked with adult development, 60 per cent had an income-generation programme for beneficiaries, 22 per cent had an agriculture project (food or community garden), and 15 per cent had helped beneficiaries gain access to formal employment (Food Bank of South Africa 2013). This is crucial in improving the livelihood of local people and building their capacity to afford food.

Although Uganda's Hunger Project food bank does not reach beneficiaries through linked organisations in this way, it could emulate the South African example by encouraging smallholder farmers to diversify their sources of livelihood. The food bank in Uganda is already providing practical skills training to farmers directly and to trainers of trainees, who are in turn expected to train other farmers in appropriate farming methods. Details of these schemes are provided below.

Uganda

Hunger Project food banks have played a significant role in improving farming and agribusiness skills among local farmers. In 2012, it was noted that over 2,538 partner farmers (1,356 women and 1,182 men) were trained in better farming methods and in agribusiness (Hunger Project-Uganda 2012:10). This report, however, does not break down the number of partner farmers who were trained per region or food bank. It is thus difficult to establish how many from Mbale district were trained or to assess the impact of the training. Nonetheless, it was generally reported that the training boosted the adoption of best farming methods in epicentre communities, including planting early maturing and high yield varieties, post-harvest handling and storage as well as collective marketing.

The Hunger Programme report indicates that a total of 596 agriculture trainers of trainees were created (Hunger Project-Uganda 2012:10). It further notes that the food banks distributed 6,970 kg. of improved seeds and 1,050 kg. of fertiliser to farmers. The type/species of seeds distributed, their source, the number of people who benefited, as well as the number of food banks that distributed the technologies, however, could not be traced and thus it is not possible to establish how the seeds were determined to be suitable for local environments. The report, nevertheless, indicated increased yields and increased incomes for local farmers as a result of the inputs.

Therefore, it can be argued that traditional food banks that provide emergency food that is donated by the public and industries cannot in any way help people achieve sustainable livelihoods and food security. However, those that have diversified their operations, like the Hunger Project and the food banks of South Africa and Delaware, can to some extent provide effective mechanisms for achieving sustainable livelihoods and food security.

It can also be concluded from the experience of community-managed rice and grain banks that establishing such a system was the best way to empower poor communities to combat food insecurity. However, food banks on their own are not effective in eradicating food insecurity, but need to be implemented alongside other programmes such as appropriate education and training, health, water and sanitation and savings and credit prgrammes.

Sustainable Livelihoods Framework

The sustainable livelihoods framework of the Institute of Development Studies (IDS) was adopted for this study. Since the study intended to explore the concept of food security and sustainable livelihoods in the context food banking, the sustainable livelihoods approach was considered appropriate. This approach to development is people-centred and cuts across disciplines. It targets eliminating poverty and strengthening local capacities to achieve sustainability and facilitates the understanding of diverse and dynamic livelihood systems.

The IDS sustainable livelihoods framework is not, however, considered an immutable blueprint. Rather, it was adapted to the study context. It offered useful insights and worked as a useful checklist in developing data collection tools and in collecting and analysing the data.

Using the livelihood framework, the author paid attention to how the food bank helped smallholder farmers to tap into and utilise a range of livelihood resources/capital assets in pursuing sustainable livelihoods. For example, under natural capital, questions related to access to and use of land, water and trees were asked. In terms of human capital, the focus was on indigenous and modern knowledge, education levels of farmers and service providers and good health of the farmers.

As to social capital, the study investigated existing networks and relationships and how farmers were tapping into them or being helped to tap into them to build their security and adaptation capabilities. Financial capital involved the farmers' cash base, access to credit, as well as access to markets. Finally, under physical capital, the attention was on food and seed storage facilities, agrochemical inputs and the mixing of enterprises to include livestock.

Attention was also paid to what the sustainable livelihood perspective could offer in terms of seed security. It was discovered that availability of and access to sufficient good quality seed represented a major challenge to food security among farmers in the study area. The livelihood perspective was thus helpful in understanding and identifying the sources of seed insecurity as well as linking these to livelihood vulnerability and resilience.

Smallholder farmers become more secure in terms of sustainable food production when they have seed security. The availability, access and quality of seed are key factors for smallholder farmers in Uganda. Sperling *et al.* (2006) present a good framework for analysing seed security for farmers (Table 8). Sperling's model has been adapted and integrated into the livelihood analysis of smallholder farmers.

According to Sperling, Remington, and Haugen (2006:3), *availability* is essentially a geographically-based parameter, and so is independent of the socio-economic status of farmers.

Seed access is a parameter specific to farmers or communities. It largely depends upon the assets of the farmer or household in question: whether they have the cash (financial capital) or social networks (social capital) to purchase or barter seed.

Seed quality includes two broad aspects: Seed quality *per se*, and varietal qual-

Table 8: Seed security framework: Basic parameters

Parameter	Seed Security
Availability	Sufficient quantity of seed of target crops within reasonable proximity (spatial availability), and in time for critical sowing periods (temporal availability).
Access	People have adequate income or other resources to purchase or barter for appropriate seeds.
Quality	Seed is of acceptable quality and of desired varieties (seed health, physiological quality, and variety integrity).

Source: Sperling, Remington, and Haugen (2006:3)

ity. Seed quality consists of physical, physiological and sanitary attributes (such as the germination rate, and the absence or presence of disease, stones, sand, broken seed or weeds). Varietal quality consists of genetic attributes, such as plant type, duration of growth cycle, seed colour and shape, palatability and so on. These parameters offered a useful checklist in investigating the food bank's role in food security, assessing the validity of responses by different respondent categories and in appropriately analysing the findings.

The study started by establishing the main sources of household food and seeds for planting. From all the six focus group discussions, two common sources of food were established. These included farming from household gardens, and buying from shops when household stocks ran out. In one group, however, some farmers mentioned a third source of food: tilling gardens for other farmers in return for food as pay (commonly referred to as *Leja Leja*).

A variety of seed sources were mentioned across all the focus groups and in individual interviews with farmers. The main sources were: 1) buying from farmers' shops such as Sukura and El Shadai; 2) saving from farm harvests; and 3) seed loans from the food bank. There was general agreement among respondents that most times the food bank supplied quality seeds, but they usually came late and in insufficient quantity. As a result, output has usually been low, affecting the rate of seed loan repayment by farmers. Thus, the accessibility and availability factors presented by Sterling *et al.* (2006) affect the output of farmers.

The two main sources of food and seeds mentioned by respondents were in line with the findings of the WFP (2013). WFP indicates that although the majority of Ugandans were involved in farming, they were also fairly market oriented, with the market contributing up to 50 per cent of household food calories. These findings also touch on the physical and financial capital elements presented in the livelihood framework (Scoones 1998). Farmers are utilising their physical capital (land) to grow the food they need for their households. Many farmers also sell most of their harvest to local markets to get the financial capital to meet other needs. The sale of farm output was thus one source of livelihood.

The other sources of seed mentioned included government programmes such as NAADS and NARO, which worked as partners of the food bank in offering services and seeds to farmers. Some rice farmers got seeds from international NGOs like the Japanese International Cooperation Agency (JICA), while others who were members of farmer groups sourced seeds from associations such as Busoba Tubana Mixed Farmers' Association (BUTUMFA). Yet others sourced them from friends and relatives. NAADS is the main instrument through which the government of Uganda is implementing food security programmes (MFEPD 2012), while NARO distributes seeds for purposes of propagation and dissemination of new technologies. The finding, however, indicates that farmers have been utilising their social capital efficiently to locate and acquire seeds from various sources.

It was evident in this study that there was a high level of relief dependency among farmers on outside sources. Farmers felt very disappointed if they did not get the amount of seeds they expected from a particular provider, such as the food bank or NAADS. It is therefore important that these farmers are trained and encouraged to save their own seeds from their harvests, which is an affordable and reliable means of achieving food security. Since the food bank is offering storage services for farmers' food, they could also take advantage of this to store their seeds for the next harvest.

The challenge, however, is that many farmers have resorted to hybrid species, which cannot be recycled more than once. Berge (1996) observed that the effect of increasing agricultural modernisation is that farmers are purchasing more of their seed and consequently reducing the role of indigenous knowledge in food security. Encouraging seed saving will permit farmers to have different varieties of each crop to allow for varied physical environments and as a coping strategy for the complex risks associated with climate variability.

Challenges to accessing sufficient food

It is believed that today smallholder farmers are experiencing a number of interlocking stressors, other than climate change and climate variability (HLPE 2012:281), which limit their access to sufficient food. This was evident among the farmers who participated in this study.

Access to sufficient quality seeds

Although smallholder farmers in Mbale indicated that they had a variety of seed sources, access to sufficient quality seeds was found to be one of the major factors affecting their food security. Farmers stressed there was a lot of duplication of the seeds sold in farm shops and even at times supplied by the other sources mentioned earlier. As a result, their ability to improve food production was jeopardised. What, therefore, can help smallholder farmers to ascertain the quality of seeds before they buy and sow them?

Ravinder et al. (2007:4) suggest that farmers need to know which seed supplies are healthy in order for them to increase their crop yields significantly. Many smallholder farmers often inspect the seed before purchasing it from farm shops or local markets, but the quality of seed is not always obvious to the naked eye. The seed security framework used in this study indicates that seed quality consists of physical, physiological and sanitary attributes (Sperling et al. 2006).

Whereas farmers may be able to observe the physical quality, they may not be in position to evaluate a seed's the physiological qualities. They may also need help in ascertaining a variety's sanitary qualities, which according to Sperling et al. include genetic attributes (Sperling et al. 2006).

Poverty, however, prevented many farmers from acquiring adequate and good quality seeds for planting, which compromised their output. The food bank as well as other service providers such as NAADS and NARO provided limited quantities of seeds. Consequently, when farmers plant less, their harvests are low, and thus less or nothing is stored in the food bank. Whatever farmers produced was consumed or sold to meet other pressing household needs, thus perpetuating food insecurity.

Small and infertile lands

Another challenge uncovered by the study was limited land, which was also becoming less fertile. Farmers could not expand production by using tractors or ox-ploughs. This challenge is attributed to the limited financial capital of small-holder farmers to enable them to enlarge their lands either through purchase or hiring. Also, because the available land is limited, it is cultivated season after season without a fallow period. High poverty makes the application of fertilisers and other agro-chemical inputs difficult. These factors, coupled with high population growth that is increasing pressure on available land, have made it difficult for farmers to achieve sustainable food security.

Other factors

Other factors include large families with many members of unproductive age, and climate changes such as drought and prolonged rain, which also lead to increased pest infestations and disease outbreaks. Also germane are the lack of appropriate food storage facilities in households; limited skills in appropriate farming methods; theft; damage to crops by stray animals; family sickness, which affects family labour and resources; surreptitious sale of crops by husbands; over reliance on single farm enterprises as well as laziness and negligence. All these were mentioned as factors limiting a household's achievement of food security.

Many of the above challenges are in line with the findings of Curtis (2013:7) and IFAD (2013:9). They argue that smallholder farmers must contend with poor quality seed and land, inadequate water supplies, basic farming equipment and poor storage facilities. When these challenges are compounded by the increasing impact of climate change and poor access to local markets, extension services and rural financial services, farmers become increasingly vulnerable to poverty, hunger and food insecurity.

Role of the food bank in food security

The food bank in Busoba is used regularly by farmers, primarily those living in close proximity. It is thought of as a necessary community resource, especially for food storage after harvest; improved seeds for planting; and technical knowledge and skills. Farmers appreciate the role the food bank has played in encouraging them to grow indigenous crops that are drought- and pest- resistant. Others appreciated the food bank's role in linking them with the Hunger Project village bank to acquire loans to finance their farming (see Table 9).

It was, however, found that low harvests resulting from droughts and poor farming methods, etc. strongly affected food storage in the food bank and consequently affected food security. Many farmers in sub-counties such as Nyondo and Lukhonge did not store their food with the food bank because of distance. They cannot afford the transport costs, despite their lack of appropriate household storage facilities. It was partly because of distance that many farmers indicated they lacked access to information about the food bank and its services. Others argued that the food bank had done little in combating food insecurity. They suggested that the food banks be decentralised to sub-counties to ease access to food storage services and information, and to reduce the costs of transport.

Alternatively, the author considers that the food bank could solve this problem by having decentralised food collection centres where villagers would deposit their food. Food bank agents and volunteers could then collect and deliver it to the food bank on their behalf. This alternative, however, offers only a short term remedy and will only be possible if the food bank and its volunteers have transport, which was lacking at the time of this study. In the long term, permanent solutions may be necessary. Useful recommendations are made in the conclusion.

On the other hand, food bank officials indicated that although they tried their best to train and sensitise farmers about better farming methods and the value of storing their food in the food bank, the adoption rate was still low. Farmers relying heavily on traditional farming methods and with poor post-harvest handling facilities were reluctant to preserve and save their seeds, preferring instead to sell off their produce early rather than storing it and selling when prices were high. This posed a challenge for the food bank's efforts to help farmers save and have enough food throughout the year.

Another major challenge the food bank has to contend with is the negative mindset among farmers. Besides having a dependence syndrome on external support for their seed requirements, the study established that smallholder farmers in the study area did not take responsibility for and proper care of the seeds they received from the food bank or the NAADS programme. Interviews with the Hunger Project's Mbale epicentre chairman and the NAADS coordinator from Busoba sub-county revealed an interesting fact about farmers in this area: many farmers planted the seeds they got from the food bank or NAADS apart from those acquired from other sources, such as the market or their own saved seed. When it came to caring for the crops, they paid more attention to the crops grown from their own seeds and took little responsibility for the crops grown from food bank or NAADS seeds.

They also noted that farmers often tagged the crops according to the seed source: "this is the Hunger Project beans or maize" and "this is NAADS maize or beans." In other words, they considered them NGO crops or government crops, and took no responsibility for them. When the yields from such crops were poor, the blame was shifted to the seed providers, even though the care of the crop was not proper. The writer observes that if this attitude among many farmers does not change, food bank and government efforts to build food security among poor farming households, to which over 80 per cent of Uganda's population belong, will continue to be frustrated.

Looking at the possible causes of such attitudes would be helpful in finding solutions to the problem. One probable reason farmers paid little attention to the food bank's and NAADS's crops was lack of farmer involvement in the decisions about the quality and type of seeds procured. Farmers indicated that the seeds provided by the food bank were sometimes of poor quality and unsuited to the local climate and soils. Farmer involvement in seed selection and procurement is likely to increase acceptance and ownership of the programme by farmers, a key factor in its sustainability.

Food bank officials acknowledge that farmers complain of difficult access to the food bank due to transport problems. The food bank officer and the epicentre chairman noted, however, that this was never a problem when they had a double cabin pickup on station. They could retrieve a farmer's produce from his village and deliver it to the food bank. The pickup was later taken back by Hunger Project-Uganda, because it was becoming costly to maintain and, instead, a motorcycle was offered. They requested the Hunger Project office to return the pickup to facilitate transport. For a short time, the food bank was planning to establish seed selection centres near farmers where they could help farmers select quality seeds for storage and later sale for better prices.

The food bank, however, faced its own financial constraints. Although farmers complained of insufficient quantities of seeds, food bank staff indicated that they could not meet the farmers' demand due to their limited financial capacity (see subsection on food bank funding and sustainability). The food bank, however, tried to supply seed to farmers through its retooling programme, whereby profits from sales in the previous season were ploughed back into the purchase of new seeds. It also acquired improved seeds from partners such as NARO. At the time of this study, the food bank had already acquired NABE 15, NABE 16 and NABE 17 beans, and other strains such as K132, from the National Agriculture Research Organization (NARO), and was ready to supply farmers as the planting season approached.

NABE 15, NABE 16 and NABE 17 are quick maturing, disease-resistant and high-yielding bean varieties that had recently been released by NARO for cultivation by farmers. These varieties, and others yet to be released, were selected by smallholders because of their higher productivity in the field compared to existing varieties, coupled with their good taste and ease of cooking (*East Africa Agribusiness* 2014).

Although the food bank supplied seed to farmers on loan, the food bank in Mbale and all other food banks operated by the Hunger Project in Uganda did not offer relief or emergency food to the hungry during times of food scarcity. It trained and encouraged farmers to produce more food instead, and to save in order to prepare themselves for times of scarcity. This was intended to ensure a sustainable food supply. However, it cannot be ruled out that some of the seeds provided by the food bank on loan ended up being eaten, since there was limited follow-up and monitoring. This is one reason for low rates of seed repayment by farmers to the food bank.

This is unlike the operations of food banks such as the food bank of Delaware (2011), the community food bank of Southern Arizona (Community Food Bank 2009), food bank of South Jersey (2014), and the Chalmers Community Service Centre in Guelph, Ontario (Chalmers Community Service Center 2014), to mention just a few. These food banks offer emergency food and do not aim to achieve food security. However, some food banks have established relationships with community gardens whereby the produce grown by them is donated to the food banks.

The above food banks also give out free clothes and household items like bedding and small appliances. They offer repair services for clothing and backpacks as well as providing social time for guests to have coffee and conversation (Board Chairman Chalmers Community Service Centre 2014). Such activities are typical of traditional food banks, and are quite different from those of the Hunger Project food banks in Uganda.

The food bank in Uganda, however, has some likeness to the increasingly common community-managed rice banks of South-East Asia as a means of addressing the seasonal food crises facing poor communities (Datta 2007). As noted above, villagers borrow rice from the bank and repay the same amount when their next crop is harvested. This is basically similar to what is practised in Ugandan food banks. The difference is that with rice banks, villagers are free to use the rice from rice banks as food during times of food scarcity or for planting. In Uganda, seeds are typically given out for planting and not for food. Nonetheless, both rice banks and food banks in Uganda are considered far cheaper than borrowing from moneylenders, but also help to reduce exploitative practices, meet specific basic needs and promote collective decision-making (Datta 2007).

From the Table 9 below it can be observed that women played an active role in food bank activities compared to men. They were more aware of the benefits derived from the food bank as well as the services offered by it. It can also be concluded that men in Busoba and Nyondo sub-counties were less involved in food bank activities and thus had little to say about the role of the food bank in food security compared to their counterparts in Lukhonge sub-county.

Generally, farmers from Lukhonge were more actively involved in food bank programmes, as witness the number of responses by both men and women. Yet

Table 9: Food bank's role in food security and Focus Group Discussion (FGD) cross-tabulation

Role of food bank	FGD Group						Total out of 6
	Busoba Women's FGD	Busoba Men's FGD	Lukhonge Women's FGD	Lukhonge Men's FGD	Nyondo Women's FGD	Nyondo Men's FGD	
Food storage services	1	1	1	1	1	0	5
Quality seeds	1	0	1	1	1	0	4
Skills training	0	0	1	1	1	0	3
Link to village bank for farm credit	1	0	1	0	1	0	3
Encouraging farmers' group formation	0	0	1	1	0	0	2
Encouraging growing of indigenous crops	1	1	1	1	1	0	5
Contributing nothing	0	0	0	0	0	1	1
Limited information about food bank services	0	0	0	0	1	1	2
Exchange visits	0	0	0	1	0	0	1
Total	4	2	6	6	6	2	26

Source: Author's fieldwork (2014)

Lukhonge is located further away from the Hunger Project epicentre than Busoba and Nyondo. This study did not identify reasons for this. However, they need to be established to ensure that food bank services benefit all potential target groups equally.

Funding and sustainability of the food bank

The food bank in Mbale-Busoba relies heavily on Hunger Project Uganda for major funding. Hunger Project Uganda is an affiliate of the Global Hunger Project, a 501 (c) (3) non-profit charitable organisation incorporated in the United States (Global Hunger Project 2012). The Hunger Project raises funds in Australia, Canada, Germany, Japan, New Zealand, Sweden, Switzerland, the Netherlands, the UK and the US.

The annual global operating budget of the Hunger Project is about $18 million. Approximately 30 per cent of revenue raised originates in the US, the bulk of which comes as unrestricted funds from individuals (Hunger Project 2013b). Hunger Project operations in developing countries also raise considerable funding from governmental and multilateral sources. The Hunger Project does not consider its investors as donors, but rather as partners and stakeholders in the success of the Hunger Project mission (Hunger Programme 2013b).

Sustainability of Mbale food bank

The sustainability strategies for the food bank at the moment are based on its operational strategies.

Revolving seed loans

Interviews with the food bank officer and the chairman of the Mbale Hunger Project Epicentre in Busoba, Mbale in February 2014 revealed that, initially, the food bank received operational funding from the Hunger Project-Uganda as a revolving loan fund. The funds were used to procure seeds and other technologies to supply to farmers in the form of loans. Later, after Hunger Programme-U stopped providing funds for seed procurement, the food bank used the funds raised from the revolving scheme to procure seeds and supply them to farmers. At the time of this study, the food bank had about two million Uganda shillings in its account from its revolving scheme, and was preparing to procure seeds (interview, 2014).

Establishing partnerships

The food bank also establishes external partnerships with other agencies for services and technology. These agencies include NARO, NAADS and Mbale district local government. At the time this research was conducted, a memorandum of understanding between Hunger Project's Mbale epicentre and the East African Seed Company was being finalised for the supply of quality seed at subsidised costs to the food bank and its farmers (Interview with the chairman Mbale epicentre, 2014).

Local community empowerment

The food bank was also empowering farmers to become self-reliant by linking them with the village bank and other savings and credit cooperatives to acquire credit. This was intended to help farmers develop the capacity to meet their own seed requirements in the near future and reduce their dependence on the food bank for seed. Moreover, the very fact that food banks are managed by local people appointed by the community is a good sign of the institution's sustainability. The Hunger Project has a food security committee that manages the affairs of the food bank, and this committee is elected by the community at a community meeting. In one way or another, community members feel a sense of ownership, an important aspect of sustainability.

Local seed production

Furthermore, the food bank was growing its own food to produce crops and seed to supply locally. At the time of this research, the food bank was in negotiations with some sub-counties for land to expand its food production. Currently, the project has very little land on which to grow its food.

Women's involvement

The food bank encourages women to become actively involved in its operation and management, because of their proven commitment and efficiency. However, higher illiteracy levels among women reduce their level of involvement. This challenge is being addressed through the promotion of adult literacy through the Hunger Project's education programme. This is also aimed at encouraging women's participation in the management of the food bank.

Women's literacy levels in the area are likely to rise faster because other NGOs, including the Foundation for Integrated Community Development Programme (FICODEP), have also recently launched a free functional adult literacy programme, in which many women have enrolled. Participation by women in the food bank is expected to enhance their social mobility and participation in community decision-making processes, thereby strengthening the operation of the food bank.

Challenges to sustainability

Crop failure

Despite the food bank's efforts to achieve sustainability, discussions revealed that consecutive crop failures due to prolonged drought posed the greatest challenge. Crop failure not only made loan seed recovery difficult, but also left households with insufficient food to meet their daily requirements.

Dependency syndrome

High relief dependency was another potential challenge: some farmers believe the seeds from the food bank were relief from the NGO, and there was no need to repay. When a new assistant project officer was appointed at the Mbale epicentre, it was reported that many farmers defaulted on their seed loan, claiming they had paid the old project officer. Since there was no evidence to prove otherwise, the food bank incurred losses. The same problem existed in the microcredit department, which at the time of this study was still battling to recover all the loans disbursed by the outgoing assistant project officer.

The food bank was trying to address this problem by creating awareness among the people of the role of the project and the obligation of borrowers to repay their seeds in order for the food bank to be sustainable. Some farmers had repaid while others still had not. In the author's opinion, the food bank could also try to address this problem through exchange visits and using well-performing food banks as model learning centres for sharing experiences of how a good food bank should be managed.

Community ownership of food bank

The concept of community ownership of the food bank is still not well established in the sub-counties where the food bank works. Many villagers are still not proud of their right to make decisions about the way seeds are procured, the proper use and maintenance of seed stocks, access to information and about the kind of information circulated. There is a need to improve information about and awareness of the food bank and its services among villagers to increase their involvement and ensure sustainability.

Generally, it is too early to tell whether food bank is self-sustaining, because it is still implementing its first strategic plan and expanding its operations. It is about two years now since the official launch of the Mbale food bank, although it began operation prior to that.

How the food bank could improve food security

To improve its role in food security, farmers urged the food bank to secure and supply enough seeds on time to allow timely planting. As Curtis (2013) has said, smallholder farmers often lack influence and power over decisions that affect them. The farmers argued that they were seldom involved in deciding which type of seeds to procure. They thus wished to be involved in such decisions, because they clearly knew what they needed based on the nature of their soils.

Farmers also suggested receiving a variety of seeds from the food bank in order to spread risks. For example, on top of receiving maize and beans, they also wanted vegetable seed. They urged the food bank to also supply fertilisers and pesticides on loan. They were willing to pay back the loans in line with the food bank's terms and conditions. Many farmers said their crops failed to yield well because they did not know the appropriate type of fertiliser and pesticides to apply. They thus wished for the food bank to supply these and give technical advice on their use.

Four of the six focus groups suggested that the food bank should be brought closer to the people. They agreed with having a main centralised food bank at the Hunger Project epicentre, but also suggested having sub-branches in sub-counties. They also wished to have information offices decentralised to the sub-counties. Farmers in Lukhonge and Nyondo were even willing to offer land for the information offices and food bank sub-branches. The demand for having food bank services brought closer highlights the important role the food bank could play in fighting food insecurity if its operations became more effective.

Whereas female farmers in Busoba and Lukhonge sub-counties wanted the food bank to supply ox-ploughs for hire, the men in the two sub-counties suggested that the food bank have a tractor that they could hire to increase the land under production. This suggestion could imply that the farmers' mindset is broadening and they were moving towards a shift from traditional household labour to mechanised farming, which would improve output. Nonetheless, many farmers still believed their mindset needed to be changed for good through training. This was mentioned in three focus group discussions and by almost all eight of the individual farmers interviewed. They also encouraged raising awareness of the services offered by the food bank.

Although farmers from Lukhonge suggested that the food bank conduct village-level skills training, this is unrealistic given the huge numbers of villages per sub-county and the food bank's limited capacity. On average, a sub-county in Uganda has about 30 villages (author's calculation based on UBOS (2002) statistics).

A possible alternative, in the author's view, would be increasing the number of farmer field schools and locating them strategically in terms of access by farmers and availability of land for demonstration. Currently, the food bank has only two field schools, one each in Lukhonge and Busoba sub-counties, and a demonstration garden near the food bank premises. This information is summarised in the Table 10.

Other key informants made suggestions similar to those made by the farmers (securing enough seeds; increasing sensitisation; supplying farmers with subsidised fertilisers and pesticides; change farmers' mindset), but also offered other useful suggestions. Five of the six key respondents suggested close monitoring of

Table 10: Farmers' suggestions for improving food security by the food bank and name of FGD cross-tabulation

Suggestion	FGD Group						Total out of 6
	Busoba Women's FGD	Busoba Men's FGD	Lukhonge Women's FGD	Lukhonge Men's FGD	Nyondo Women's FGD	Nyondo Men's FGD	
Bring food bank near to the people	0	0	1	1	1	1	4
Supply enough seeds on time	1	1	1	1	1	0	5
Provide pesticides and fertilisers on loan	1	0	1	1	1	0	4
Provide ox ploughs for hire	1	0	1	0	0	0	2
Train farmers in improved farming methods	0	0	0	1	1	1	3
Conduct village-level farmer training	0	0	0	1	0	0	1
Acquire tractor for hire	0	1	0	1	0	0	2
Train to change farmers' mindset	0	0	1	1	0	1	3
Provide transport to food bank	0	0	0	0	1	0	1
Involve farmers in deciding which seeds to procure	0	0	1	0	0	1	2
Sensitisation and creating awareness of the food bank	1	0	1	0	0	1	3
Total	4	2	7	7	5	5	30

Source: Author's fieldwork (2014)

and follow-up on farmers who received seeds and training from the food bank, to ensure they were doing things right. Both farmers and key informants felt this was lacking in food bank operations.

One possible reason for this is that the food bank was mainly run by volunteers, few in number, and with many other commitments. The Hunger Project seeks to empower local people to manage their own development through beneficiary participation. Nonetheless, reliance on volunteers could also result from limited funding for the food bank, but also as a strategy to build local capacity as well as ownership to ensure sustainability. It can also be observed that almost all food banks across the world use volunteers to run their activities (Agostinho and Arminda 2012).

Therefore, it is important that food bank personnel understand what motivates these individuals to volunteer their valuable time for food bank service and activities, in order to achieve better performance (Brand *et al.* 2008).

Key informants suggested that the food bank build the capacity of its staff and volunteers to deal with food security issues among smallholder farmers. This could be achieved through training and research on seed preservation and value addition. The informants encouraged the food bank to conduct more research on how best to preserve farmers' food as well as to add value to it. This was expected to improve the prices obtained for produce as well as farmers' livelihoods. However, given that the food bank in Uganda, as elsewhere, often employs voluntary labour and has very low key fundraising (Husbands 1999; Agostinho and Arminda 2012), it lacks the human and financial resources to undertake research and innovation to systematically address food insecurity and hunger.

The food bank could, however, cooperate with agricultural colleges and universities (for example the Arapai and Bukalasa agricultural colleges, Makerere University College of Agriculture and Environment Science) in conducting research and documenting collections; controlling the quality of seeds; as well as multiplying the seed varieties of interest to farmers (Development Fund, Norway 2011). Other food banks, like those in Southern Arizona (2009) and Delaware (2011), cooperate with universities and make use of agriculture students as skilled manpower through internship programmes. Table 11 displays the recommendations by key informants.

Despite the attempts made to combat food insecurity, the food bank still faces challenges like the negative mindset of farmers towards new farming methods and post-harvest handling problems. Nonetheless, there was clear evidence the food bank had great potential to help farmers improve their food security. Farmers, project staff as well as key informants all indicated that the food bank trained and encouraged farmers to grow indigenous crops that were drought- and pest-resistant; plant improved speciesand apply fertilisers and manure; and

Table 11: Key informants' suggestions for improved food security by food banks

Suggestion	Frequency	Percentage
Secure enough seeds	4	12.5
Research on seed preservation and value addition	6	18.8
Capacity building of food bank staff and volunteers	4	12.5
Increase sensitisation and knowledge about the food bank	3	9.4
Change farmers' mindset	5	15.6
Offer seeds, fertilisers and pesticides at subsidised prices	5	15.6
Total	**32**	**100**

Source: Author's fieldwork (2014)

grow both cash and food crops alongside rearing animals and poultry. Generally, farmers who adopted these practices achieved much better food security than those who did not.

Food bank as mechanism for achieving sustainable livelihoods

Sustainable food security in developing countries such as Uganda, in which the majority of families derive their livelihoods from agriculture, cannot be pursued in isolation from sustainable livelihoods. Can the food bank be a mechanism for achieving sustainable livelihoods among smallholder farmers? That is the question.

The issues of poverty reduction and environmental management have been central in the debate on sustainable livelihoods (Scoones 1998:3), as has building resilience to climate change. This section is divided into two main subsections: Livelihood resources and strategies; and building resilience to climate change.

Livelihood resources and strategies

Exploring the institutional (food bank) influences
Both farmers and food bank officials acknowledged that the food bank helped farmers gain access to the credit they used in their farming. For example, before a farmer qualified for a seed loan from the food bank, he /she had to be a member of a registered farmer group. Farmer groups were able to access group loans from the Hunger Project village bank, sometimes upon the recommendation of food bank officials, often upon assessment by the loan committee. This not only supported farming activities, but also enabled investment in other business activities to diversify income sources. Women of the Lukhonge sub-county focus group joyfully asserted:

> We got a loan from the hunger project village bank, invested it in a business and make much profit. Now we have been able to open up our own food store and other farmers keep their food with us (Women FGDLukhonge 2014).

This suggests the sustainability of the food bank programme, and such initiatives should be encouraged to reduce the costs of managing food banks at centralised locations as well as to encourage farmers to save food for food security.

Farmers indicated that their employable time had increased, as they were occupied in many farming enterprises and non-farm enterprises like petty trade, because their farm incomes had improved. When asked whether the food bank had helped them improve their incomes, they unanimously said yes, except for the Nyondo men's FGD, which indicated it had got no help in that regard.

Farmers maintained that the food bank had provided storage facilities that enabled them to store their food until market prices were high, while keeping the rest for future household consumption and planting. They said this had been difficult in the past because of their poor storage facilities, which compelled them to sell their harvests early to middlemen for paltry prices.

Others said that with food easily accessible in their houses, it was susceptible to waste from rodents or pests, or was sold by husbands for alcohol, thus affecting household incomes. They appreciated that in the food bank, their food was secure from man, pests and rodents.

I observed, however, that although the food bank was rodent free and offered security for farmers' food, it lacked proper stalls for keeping the produce. Food bags were laid on the ground or on low tables, and could easily be invaded by grain beetles and other pantry pests. The researcher recommends that better, high enough stalls be put in place to reduce the risk of pests and resultant losses to farmers.

Besides teaching better farming methods, in itself a strategy to help farmers boost their incomes, the food bank encouraged farmers to form savings and credit cooperatives (SACCOs). SACCOs are a network of cooperatives, ranging from community-based initiatives, to recruiting members from the informal economy, to workplace-based groups (*Cooperative Societies Act* 2003). People come together in cooperatives save money, take loans from within their group as well as get funding from government and commercial financial institutions to support their activities. SACCOs in Uganda are regulated by the Uganda Savings and Credit Cooperation Union (UCSCU 2014).

The food bank's support for the formation of SACCOs was by encouraging farmers to develop a savings culture and reduce overspending of their hard earned incomes. Many farmers interviewed did not have accounts with commercial banks. They kept all their money at home, thereby risking unnecessary spending and theft. Forming and joining saving groups offers an affordable means of saving money and getting loans whenever necessary. Some farmers had already joined the Hunger Project village bank, and were proud of their increased savings. One female respondent exclaimed: "I was shocked when I took a financial statement of my account with the Hunger Project village bank; my account had almost a million shillings from savings. It is the first time I've saved money" (Lukhonge Women's FGD 2014)

Food bank officials also realised that many farming households spent most of their incomes on treating common illnesses. Through their change agents, the food bank started to sensitise farmers about household sanitation and hygiene to reduce the incidence of common illness. They also encouraged proper nutrition for children and other household members. This, in one way or another, reduces household expenditure, and allows for more active labour time, which boosts household incomes.

Besides referring community members to the Hunger Project health centre in the epicentre building, community change agents (animators) actively sensitise communities about how to prevent the spread of malaria, HIV/AIDS, maternal mortality as well as sanitary-related illnesses (interview with the Hunger Project chairman, Mbale and animators, 2014). This activity falls under the Hunger Project's nutrition and healthcare programme (Hunger Project-U 2013). Community change agents are trained in various community health skills in partnership with the government health department, before being sent out to help communities.

This form of service delivery is necessary, especially in a health system such as Uganda's: despite improved access to maternal and child care as well as responses to HIV/AIDS, service delivery to address infant and maternal mortality is still poor (Ministry of Health 2012). Primary healthcare remains difficult of access for some, and the quality of care is inconsistent. The referral system is not functional, and patients often ignore secondary or tertiary care due to the high costs involved (Ministry of Health 2012). The role of community animators as health workers is therefore significant in cases where formal health services are inadequate.

Mixed enterprise strategy

The food bank trained farmers in adopting a mixed enterprise approach to boost their household livelihoods and food security. This project was appreciated by famers. The food bank encouraged farmers to grow cash crops like coffee and raise dairy cows, pigs, improved goats and poultry alongside crop growing. The food bank also has a local poultry demonstration project as well as a piggery, which farmers visited to learn how to run a mixed farm enterprise on a small plot.

Lack of a constant water supply hampered the food bank in the running of its demonstration farm. The project used water supplied by the national water and sewerage corporation, and water was pumped from a small tank at the demonstration garden. Because of the high water bills in relation to demonstration garden output, especially the vegetable gardens which required constant watering, the plots were often abandoned during dry season.

Nonetheless, the maize demonstration farm had the best stands of maize in the area, despite the scorching sun at the time of this study, largely thanks to the good farming methods employed. The banana demonstration garden also had good bunches, despite the extreme weather conditions at the time.

It was observed that farmers who had adopted the mixed enterprise strategy had better results and higher chances of achieving sustainable livelihoods and food security than those who did not. One model farmer in Busoba sub-county practising mixed farming is a good example. Manure and urine from his dairy cows was used in the coffee and banana plantations. He also interplanted coffee with trees, a practice that helped conserve the environment, but also generated income for his household. He said:

> I have planted large acres of trees from which I cut timber once in a while. I have inter-planted trees with coffee from which I have been able to educate my children and many of my siblings. I have dairy cows which served me milk throughout the year for family consumption as well as selling to meet petty family needs. The animals also supply me with manure and urea that I use on my food crops and cash crop plantations. I am now a model in championing the growing of indigenous crops. I rarely buy food in my home, except for a few things like sugar, salt and cooking oil I cannot produce locally. (Individual interview 2014)

Other farmers that adopted this strategy celebrated its benefits in driving them towards sustainable livelihoods and food security. The food bank operated a coffee nursery project from which mixed enterprise farmers could purchase quality seedlings for planting at an affordable 100 shillings per seedling, half the prevailing market price.

Marketing

Although some project staff and farmers indicated that the food bank linked farmers to better markets for their produce, many farmers and informants indicated that these efforts were still lacking. As much as such services were in the food bank's plans, farmers tended to market their produce individually.

Among the reasons food bank officials gave for not being able to find appropriate markets was the low output of many farmers, which did not necessitate collective marketing. Second, some farmers sold their produce while it was still in the ground. Third, other farmers did not select their seeds well, making marketing difficult. Nonetheless, some of the farmers appreciated the food bank's linking them with schools to buy their produce. At one time, the food bank had also connected farmers with the WFP.

Table 12: Farmers' suggestions on how food bank can help achieve sustainable livelihoods

Suggestion	Count
Search for good markets	2
Emphasise enterprise mix	4
Introduce vocational skills training for youth and women	1
Educate farmers in food and money-saving skills	2
Supply farmers with variety of seeds	4
Supply fertilisers, pesticides and pumps	4
Lobby village bank to reduce interest on loans	3
Hire or employ agricultural extension workers	3
Value addition of stored food	1
Total	**24**

Source: Author's field work (2014)

How to improve the food bank's role in sustainable livelihoods

Farmers made several suggestions about how the food bank could help them im-prove their livelihoods. Table 12 summarises the suggestions of the six focus groups.

When asked whether the food bank could be an effective mechanism in helping smallholder farmers achieve sustainable livelihoods, all key informants answered emphatically that it could. They argued that food storage facilities remained the major challenge for many farmers in Mbale. If the food bank increased awareness of its services, taught farmers better farming methods to improve yields, and increased farmers' skills in saving seed to reduce reliance on purchases and handouts, it could without doubt be effective in driving farmers towards sustainable livelihoods.

Unlike the conventional food banks studied by Teron and Tarasuk (1999) and Husbands (1999) and others, whose operation largely depended upon the quality and quantity of donations from the public, food producers, processors and retailers, the Hunger Project food bank had effective measures in place to move smallholder farmers towards sustainable livelihoods. Farmers could store their food at the food bank and sell it when prices were high and were linked to financial services through the village bank. The mixed enterprise strategy diver-sified and improved farmers' sources of livelihood.

This study therefore recommends that agricultural production activities with forward and backward linkages be promoted, as they can give rise to an array of possible agribusiness activities for sustainable rural livelihoods. The food bank, however, needs to help farmers to develop and standardise the management tools for these enterprises to suit the requirements of smallholder farming units.

Building resilience to climate change

The effects of climate change are already being felt globally, although many people are still ill-prepared for the risks associated with it (UNEP 2014). In

2007, the International Climate Change Risk Report labelled Uganda as one of the most unprepared and most vulnerable countries in the world (CIGI 2007). Vulnerability is a concept commonly used in relation to appraising effective responses to climate change. It covers a variety of concepts and elements, including sensitivity or susceptibility to harm and lack of capacity to cope and adapt (IPCC 2014b:4).

Uganda's high vulnerability is a result of its heavy dependence on primary production and natural resource use, along with rapid population growth, weak institutional capacity, limited financial resources and low income per capita, limited healthcare and economic infrastructure, limited capacity and equipment for disaster management, and heavy reliance on subsistence, rain-fed agriculture (MWLE 2002; MoFPED 2004).

Climate change in Uganda is manifested in different ways. Global climate change models project that Uganda will experience an increase in average temperatures by up to 1.5° C in the next 20 years (GOU 2009) and that Uganda's temperatures will be 4.3° C higher by 2080 (IPCC 2007). This is expected to increase levels of poverty and food insecurity (World Bank 2013). Today, the intensity and frequency of drought in almost all regions of the country is clear evidence of climate change.

The long April-August 2013 drought left a number of people hungry in eastern Uganda (Republic of Uganda 2013). Floods have also become common of late. For example, in 2007 the Teso region received the heaviest rainfall per annum in 35 years. Some 50,000 households were affected and crops destroyed, leading to food insecurity (WFP 2008). The heavy rains that lasted for six hours on 1May 2013 in Kasese, western Uganda, led to the River Nyamwamba bursting its banks and flooding in nine sub-counties. The Uganda Red Cross Society (2013) reported that eight people were confirmed dead and at least five people were missing. Some 25,455 people were severely affected as a result of the destruction of houses, crops and facilities such as bridges, roads, hospitals and power lines.

Heavy rains have also been associated with landslides, mainly accelerated by human activity, as well as the loss of life, property and financial resources (Ministry of State for Relief, Disaster Preparedness and Refugees 2010). Prolonged droughts and frequent floods in Uganda have resulted in increased pests and diseases in plants, livestock and humans. For example, the outbreak of moths in eastern and central Uganda in March 2012 (Republic of Uganda 2013) was followed by an invasion of dangerous Looper caterpillars in Masaka, Buikwe, Mukono, Buvuma, Rakai, Wakiso and Bulambuli districts in April (*Daily Monitor*, April 2012). These and many other pests have caused much destruction of farm lands, further necessitating the devising of means to address the effects of climate change in Uganda.

Farmers in Uganda need to be assisted in adapting to reduce the risks from a changing climate.

Adaptation by smallholder farmers to climate change in Uganda

This study established that although they were already being affected by climate change, farmers were not familiar with the concept. Smallholder farmers in Uganda, as elsewhere in Africa (Diner *et al.* 2008), have adapted to climate change in various ways.

Although smallholder farmers had less control over nature and sometimes could only pray to God for rain, many of them took other locally affordable steps. Farmers with plots of marshland and alongside springs grew crops during the dry season, as they could easily water them. Many swamps and springs, however, dried up during prolonged dry spells.

Some farmers mulched their gardens using plant residues to protect them from direct sun. They also applied fertilisers, organic manure and other biodegradable matter to improve soil fertility and maintain soil structure.

Spraying against pests was also practised by farmers who could afford the pesticides, while others uprooted affected plants and burned them. It was also established that some farmers made their own pesticides by mixing herbs, red pepper, ash and animal urine. The mixture is left to ferment for 3–4 days, then diluted with water and sprayed on the crops. It was reported that the mixture was effective against a variety of field pests and diseases, as well as adding urea to the soil, thereby improving yields.

Farmers also changed their planting season based on the availability of rainfall, and timely and proper weeding. They often received information about the expected arrival of rains from NARO through the food bank and planned accordingly. To prevent storage pests, all farmers reported they dried their cereals thoroughly before storing, while those that could afford to also applied pesticides.

Those who could not afford to spray stored cereals and legumes in a mixture of chaff before sieving. This method was reported to be effective in protecting stored crops against pests. No farmers irrigated, although they learned about irrigation during food bank training. Even at the farmer's field school in Lukhonge for example, the crops of farmers who did not irrigate were drying up at the time of this study, while those of farmers that irrigated on the same demonstration farm were doing very well.

These findings are in line with UNEP's (2011): many farmers in Uganda depend heavily on rainfall and use fewer inputs like fertilisers. This leaves these farmers vulnerable to changing rainfall patterns and the climatic variability associated with global warming. Moreover, Uganda lacks precise rainfall data. Thus, as a country Uganda is not climate resilient.

Food bank strategy for building resilience to climate change

Soil preservation and weather information

The food bank helps farmers with proper soil conservation and rainfall predictions through other service providers such as NAADSand NARO. The latter in turn received help from UNEP to develop efficient systems of collecting, recording and analysing agro-meteorological information, such as rainfall amounts and patterns (UNEP 2011:23). This enabled NARO to strengthen its knowledge of the uncertainties faced by farmers and provides them with information on expected rainfall and amounts.

However, a senior meteorologist at the Ministry of Water and Environment, Khalid Muwembe, noted that the projections could not be relied upon unreservedly on account of increased climate variability. He indicated that previously there had been two good planting seasons, March to May and September to November, which were reliable and allowed farmers to follow traditional planting patterns (*East African Agribusiness* 2014). However, this situation has changed, because occasionally there is continuous rainfall during the dry season and prolonged dry spells during rainy seasons, making it difficult for farmers to plan well. He therefore advised farmers to seek expert advice before planting.

Indigenous crops

The food bank also encouraged farmers to grow drought-resistant, indigenous crops, as well as encouraging the early preparation of gardens during the dry season and planting after first rains. As mentioned earlier, the prolonged droughts and floods in Uganda have led to infestations of pests that affect crops on many farms. In preparation for this, the food bank uses the services of agricultural extension assistants to recommend appropriate pesticides to farmers. The food bank also at times supplied pesticides and spray pumps to farmers on credit.

Dry-farming

Promoting proper soil conservation and indigenous crops may imply that the food bank is trying to build resilience to climate change through dry-farming. This practice requires farmers to prepare their soils well by deep ploughing, cultivation and allowing for the planting of crops. This should, however, be accompanied by the choice of crop, its proper seedlings and its proper care and harvesting.

Several crops have been considered appropriate for dry-farming and the food bank has been encouraging some of them. It is hoped that if farmers adopt this kind of farming, they will build some resilience to climate variations. The crops the food bank is emphasising for this purpose include climbing yams (or air potatoes). This is a herbaceous, high climbing plant with vines of up to 65 feet long. It climbs up shrubs and trees and produces aerial tubers (bulbils), one to four occurring at leaf axils, which drop and sprout to form new plants. The fruits usually start yielding between June and September and year-round. Climbing yams are resistant to drought and can survive all year-round under any weather conditions.

Sweet potatoes are also suitable for dry-farming, and give good yields even with rainfall of less than 12 inches. They are also well adapted for rotation and can thus be used for land fallowing. Given that they are cover crops, potatoes control ground moisture and their growth should be encouraged, especially when low rainfalls are predicted.

Other dry-farming crops advocated by the food bank include sorghum, which promises to be a good yielder even under arid conditions, although many farmers in Mbale have not grown it of late.

From the author's experience of the area, practising dry-farming in Uganda and Mbale, where the land is becoming less fertile each year, will not yield better results without inter-planting with leguminous crops. Such crops will be of great importance to farmers, because they are multipurpose. First, they are valuable animal feeds, and second, legumes have the ability to gather nitrogen from the air and can be used to maintain soil fertility. Research, however, shows that the type of legume and inter-crop chosen as well as the spatial requirement play an important role in determining yields.

According to Ghosh (2004), spacing of plants, planting rates and maturity dates must be considered when planning intercrops. Legumes play a proven role in suppressing weeds in particular crops. For example, intercropping soybean and sesame and sorghum and cotton significantly decreases the biomass and density of weeds and increases net return (Iqbal *et al.* 2007). Malik *et al.* (2008) also reported that wild radish and rye cover crops reduced total weed density by 35 and 50 per cent respectively, without herbicides.

It can therefore be argued that intercropping will not only increase soil nutrient levels, but also control weeds and in some cases reduce herbicide costs. Legumes improve soil fertility through nitrogen fixation. While recycled plant residues and animal manure help to maintain the overall nutrient balance on the farm, the single most important source of nitrogen is atmospheric fixation by legumes (Briggs *et al.* 2005).

Nitrogen is supplied in legumes from root nodules in which *Rhizobium spp.* bacteria convert atmospheric nitrogen into a form usable by the plant. The amount of nitrogen fixed in the roots of grain legumes has been estimated at 150-200 kg./ha., most of which is removed in the grain of the crop (Fisher 1996).

One mistake many smallholder farmers make is removing crop residues from the field during harvesting. Farmers uproot legumes such as beans, soya and groundnuts and take everything home; residues are often burned. Jensen (2002) argues that this practice removes much of the fixed nitrogen in the soil and reduces the benefit to following crops. It may also result in nitrate (NO_3) leaching, especially in environments which receive surplus precipitation. Jensen (2002) suggests this situation can be mitigated by early sowing of subsequent cereals or by nitrogen-catch cropping

Afforestation

For purposes of air purification and influencing rainfall, the Hunger Project food bank also promotes tree planting. It encourages farmers, even those with limited land, to plant trees along the edges of their gardens. The project has also managed a tree nursery project, from which farmers can get a variety of seedlings at low cost. This project actively involves the youth in environmental protection as well as to earn a living.

Farmers have adopted this strategy, although adoption is still low. Some complain that their small lands cannot support both crops and trees, while others argue that some trees deplete soil moisture, vital to crops. However, those who have interplanted trees like *Calliandra calothyrsus* say there have been no such effects on crops. Instead, they make the soil more fertile. Some youths have been inspired to set up their own tree nursery projects, and report earning good money each year.

Irrigation facilities

The food bank also has a drip irrigation system on its demonstration farm. Here, farmers are trained in vegetable growing and management during the dry season. As already noted, the system is expensive on account of its reliance on water from the national water and sewerage corporation grid. Such systems are difficult for farmers to adopt, because the area has limited water sources. Indeed, its adoption by the food bank itself was not cost effective. The vegetables grown on the demonstration farm shrivelled because the food bank could not afford the water bill, as output did not cover production costs.

This researcher recommends that rainwater harvesting be considered as an alternative. Underground tanks should be installed to collect rain water from the Hunger Project 'L' building, including the food bank structure, to supply cheap and sufficient water for irrigation.

In a nutshell, the study established that helping farmers to meet the climate challenge calls for individuals and institutions to be able to assess and understand climate change, design and implement policies and, most importantly, take action on climate-resilient growth (UNEP 2011:5). The food bank may therefore need to partner with other service providers to help farmers adapt to an agro-environment zoning strategy. This enables assessment of land availability and suitability, indicating which crops are appropriate to specific areas and identifying exclusion zones to protect areas of high biodiversity and carbon storage value (UNEP 2011:11).

The aim of this research was to identify the roles food banks play in ensuring food security, livelihood sustainability and in building smallholder resilience to climate change. Similar studies have been conducted in developed countries, specifically the US and Canada, with a focus on food security and the health of food bank users, but no similar study has been conducted in Uganda. Furthermore, it has been widely noted that hunger and food insecurity is increasing the need for food banks and emergency food providers and that climate change in Uganda is very likely to exacerbate food insecurity. Yet no study has been conducted to establish how food banks could be useful in building food security, sustainable livelihoods and climate resilience for over 80 per cent of Uganda's population.

Previous literature was reviewed and a starting point established for analysing the role of food banks. Three main research questions were formulated to find out the extent to which food banks helped address food insecurity among smallholder farmers in Uganda; whether food banks could be a means for moving vulnerable smallholder farmers to a state of sustainable livelihood; and whether they could help them build resilience to climate change shocks.

The findings revealed that currently it is hard to measure the socioeconomic impact of the food bank in Mbale on smallholder farmers, because it is difficult to isolate its contribution from those of related programmes such as NAADS; farmer associations such as the Busoba Tubaana Mixed Farmers' Association; other government programmes, including free primary and secondary schooling; and the other activities farmers engaged in to generate income. However, judging from discussions with farmers and key informants, it can confidently be concluded that the food bank was playing a significant role in improving smallholders' food production and income levels in the sub-counties where it was operating. Since poor farmers can now apply the better farming methods and improved seeds they acquire from the food bank, as well as store their harvest at the food bank, their daily earnings have significantly improved.

It was also found out that much of the relevant literature indicates that food banks exist to provide emergency food to the hungry and not ensure food security and sustainable livelihoods for farmers. Food banks collect would-be wasted food from industries, supermarkets and other sources and either distribute it directly to the hungry or through frontline agencies. "Conventional" food banks thus serve two main goals: assisting low-income consumers in gaining access to food and distributing surplus food. The objectives are not to end hunger or promote food security.

This study revealed that the operation of the food bank in Uganda was very different from the situation in Canada and the US. In Uganda, the overall goal of the food bank *is* to end hunger and starvation which, when achieved, is tantamount to food security. The food bank in Uganda does not offer emergency food relief to the hungry. Instead, it encourages the growing of more food and saving for the future so that farmers have greater security in times of famine and food scarcity. It also provides quality seeds to farmers and safe food storage facilities, and offers skilled training to farmers to improve their farming knowledge and yields. It also organises farmer field schools and demonstration plots for hands-on training.

Not only that, the food bank was also found to be active in helping farmers diversify their sources of household income through mixed enterprises, by linking them to credit services, as well as by helping them to become involved in savings and credit cooperatives. There is evidence that farmers who have followed the food bank training programmes have improved their income levels, others are already food secure and yet others are on their way to food security. Nonetheless, the food bank found it difficult to persuade all farmers to adopt the new farming methods and technology, either because of poverty, negative attitudes to change and the attitude of dependence on external help.

In light of all this, the author finds the food bank hunger-alleviation model suggested by Husbands (1999) less applicable to food banks in Uganda. The model would be more relevant if the food bank were dealing with urban dwellers not directly involved in food production. Since the food bank in Uganda aims to work with rural smallholder farmers, this study finds the above model less relevant and thus recommends modification.

The hunger alleviation model looks at food bank users only as recipients, not contributors. They are only at the receiving end, making them more vulnerable and food insecure, especially if the food bank is unable to supply food. Using the Hunger Project food bank project in Uganda, which works in rural areas directly with farmers, the author proposes a new hunger-alleviation model adapted to smallholder farmers. The new model would look at food bank users as the key providers of food to the food bank, and the central focus. Smallholders would turn to the food bank for knowledge, skills and other inputs, and then produce food that they store with the food bank, rather than looking to the food bank for emergency food donated by industries and the public.

It can be deduced from this study that while smallholder farmers turn to the food bank for seeds and skills/technical advice, they also have direct access to the same seed suppliers and service providers as the food bank. Those with money can get their seed directly from the seed suppliers or from NAADS. Smallholder farmers can also buy seed from the market, while later supplying their produce directly to the market. They can also store their food at the food bank after harvest, and the food bank can help to find markets, involving either urban dwellers or larger entities such as the World Food Programme.

It is also clear that the food bank's seed suppliers also supply seeds to the general market. From the above model it can be observed that the market is the convergence point for almost all stakeholders involved in food security – the food bank, urban dwellers, farmers and seed suppliers – hence corroborating the fact that Ugandans are market oriented. Although food bank partners/service providers are not directly connected to the market in the above model, in one way or other they do have a connection to the food market.

Contribution of the study

This is the first study of the role of food banks in food security, sustainable livelihoods and climate change resilience among smallholder farmers in Uganda. It has been conducted at a time when food security and livelihood sustainability is at the forefront of Uganda's development agenda. It was also timely in that the concept of food banks is still new in Uganda, and they are often confused with food barns or granaries.

This research contributes to the analysis of strategies to build food security, sustainable livelihoods and climate resilience among smallholder farmers. More specifically, the empirical findings contribute to an understanding of the need to establish community-based food security systems, whereby the capacity of smallholder farmers to steer food security in a sustainable manner is rejuvenated.

Implications of the study

This research has yielded useful insights into food security and smallholder agriculture, especially in Uganda. It provides a clear understanding for service providers to smallholder farmers, which can help them adapt to changing agricultural production as a result of climate change. Improving smallholder productivity depends on gathering useful information about changing trends in farming and on making calculated decisions about smallholder preferences and choices and so to avoid imposing inappropriate technology and methods on them.

Additionally, the previous chapter shows that while some smallholder farmers were still rigid, many were flexible and already adopting new technologies and farming methods as well as adapting to climate variability in different ways. This holds out the promise that smallholder farmers can achieve food security and sustainable livelihoods, especially if the food bank and other service providers improve their decision-making through proper feasibility studies and by involving smallholder farmers in decisions.

Furthermore, the findings reveal that proper and safe food storage not only ensures that poor households have access to food for longer periods, including times of scarcity, but also improves household incomes, and hence livelihood security. The food bank therefore needs to review its strategies to overcome the problem of distance that barred many farmers from learning about and accessing its services, including its better food storage services.

Finally, the managerial and administrative implications of this study will help the food bank and other food security service providers like NAADS to revisit their strategies to improve the capacity of smallholder farmers to make decisions relating to the management of the food bank and food security, as well as to encourage community ownership of the food bank project.

Limitations of study

The current study has some limitations that should be addressed in future research. Firstly, it focuses on only one food bank, the Hunger Project's Mbale epicentre in the Mt. Elgon ecological zone. Future research should expand coverage to include different locations and climate zones. Furthermore, this research focused mainly on smallholder crop farmers. It might be useful in future to consider both smallholder crop farmers and smallholder livestock farmers, since many farmers engaged in both activities.

Additionally, this research employed qualitative methods. It might be of interest to conduct wider quantitative surveys that provide greater understanding of the practices of food banks and their role in food security, sustainability and climate change adaptation in Uganda.

Finally, this study paid attention to the role food banks played in food security, without attempting to identify what the food bank considered to be the causes of hunger so as to enable it to assess the extent to which it was addressing root causes and not symptoms. It will be of great interest for future researchers to study the causes of hunger among smallholder farmers and how the food bank could help them not only achieve food security but also food sovereignty.

In a nutshell, smallholder farmers are still food insecure and both government and nongovernmental agencies are needed to support the food security programme in Uganda. Farmers still have a major challenge in accessing sufficient quality seed to improve their production, although the food bank, NARO and NAADS are playing a significant role in this regard and in capacity building among farmers. Some farmers have adapted to new farming systems and put into practice the skills acquired from the food bank and other service providers. They have registered better yields and incomes. However, this is still small fraction and the food bank and others need to widen the scope of their activities.

Recommendations

Establish community-managed food banks

This study established that farmers wanted food banks decentralised to lower levels. This is an indication that farmers are more interested in having a community-managed food bank in their own communities, which they can easily access, and whose day-to-day decision making they can influence. It is recommended that the Hunger Project consider initiating community-managed food banks at village level, and help community members to form management committees among themselves. This will not only reduce the Hunger Project's management costs, but could also increase a sense of community ownership of the food bank movement through community involvement in decision-making.

The current centralised food bank could play a supervisory and capacity building role, as well as in helping farmers establish simple tools and procedures for the management of food banks. This step is intended to build the capacity of rural people to initiate and manage their own development. LaFond and Brown (2003:7) argue that development capacity represents the potential for using resources effectively and maintaining performance gains with gradually reduced levels of external support.

The best approach to building community capacity for development, especially among smallholder farmers in eastern Uganda and Uganda as a whole, will be to make suitable efforts and investments in training, adult education and cross-visits, so that those responsible for managing food banks can learn from one another (Oya 2001).

Equitable distribution of benefits

In discussions with the farmers the food bank served, the point was made that the benefits of the food bank needed to be equitably distributed among community members. While farmers in Lukhonge praised the food bank for the great improvement it had brought to their agricultural production and household incomes, farmers in Nyondo and some from Busoba indicated they had not benefited much. Others did not know exactly what happened at the food bank. Villagers were more aware of the Hunger Project's micro-credit than the food bank, because the microfinance department seemed more vibrant in the way it operated. This study recommends establishment of a service delivery system that benefits all targeted beneficiaries equally. Equity depends on getting community members to agree that the arrangement is fair (Mansuri and Rao 2004).

Encourage seed saving

One of the main challenges both farmers and the food bank faced was insufficient seeds for planting. Farmers usually relied on relief sources like the food bank, NAADS and others for their seed requirements. Because of their poverty, many smallholder farmers could not afford to meet this requirement. This study makes two recommendations in this regard.

First, the food bank should encourage seed saving among farmers and preferably preservation of indigenous species. It is obvious that there is a high market demand in Uganda for indigenous seeds and cereals due to dietary preferences and tastes. However, due to their rapid displacement from the market by modern seeds, farmer preference is shifting towards genetically modified species. Helping farmers to save seeds for the next planting season as well as helping them to grow traditional high value crops will enhance the capacity of smallholder farmers to become food secure.

Initiate a community-supported agriculture programme

Alternatively, the food bank could initiate and pilot the concept of community-supported agriculture. This ensures a partnership between farmers and consumers, in which the responsibility and rewards of farming are shared. Farmers could receive funds from potential consumers in advance to buy seeds, pesticides and fertilisers and prepare the land. A portion of the harvest is then paid to the consumers who provided the funds according to established terms and conditions.

Encouraging community-supported agriculture will provide multiple benefits to smallholder farmers, the agricultural sector and the economy of Uganda. First, the system ensures that money remains in circulation within the local community leading to local development, because the consumers are usually from the local community. Second, it curbs the buying of imported foods or foods transported great distances, with benefits for the environment and in terms of global warming. Third, it provides a ready market and security for farmers' produce. Finally, it protects farmers against exploitation by private money lenders and other financial institutions that charge high interest on loans.

Rainwater harvesting facilities

This study also established that smallholder farmers learn better farming methods best through demonstration and hands-on training. However, the food bank was constrained by lack of water for irrigation at demonstration sites during the dry season, rendering the demonstration farms less efficient. Reliance on the national water and sewerage corporation for water was not the best way to support the irrigation programme. This study therefore recommends that the food bank invest in rainwater harvesting to collect water during rainy season and use it for irrigation during dry season. Underground tanks with higher capacities would be the most appropriate strategy to adopt.

The Mbale district lies in the Mount Elgon climatic zone and experiences its main rains between March and September, with an average rainfall of between 1,250 and 1,500 mm, which is enough to support irrigation. Proposals could also be developed for funding from various sources to support a rain-water harvesting project for smallholder farmers, for example from the African Water Facility of the African Development Bank (AfDB) and the Water and Sanitation Programme-Africa Region (WSP-AF).

REFERENCES

Academy of Development Science (n.d.) About Academy of Development Science: GRAIN BANKS. http://www.rainforestinfo.org.au/katkari/ADS.html#top, accessed 20/05/2014.

ACC/SCN (1991) Some options for improving nutrition in the 90s, *SCN News* 7, Geneva. http://unscn.org/layout/modules/resources/files/scnnews7_supplement.pdf, accessed 02/05/2014.

Agostinho, D. and P. Arminda (2012) Analysis of the motivations, generativity and demographics of the food bank volunteer. *International Journal of Nonprofit and Voluntary Sector Marketing* 17:249–61. doi: 10.1002/nvsm. 1427

Alamgir, M. and P. Arora (1991) *Providing Food Security for All.* New York: New York University Press, International Fund for Agriculture Development.

Altieri, M.A. and C.I. Nicholls (2005) *Agro-ecology and the search for a truly sustainable Agriculture.* 1st ed. United Nations Environment Programme.

Anderson, S.A. (ed.) (1990) Core indicators of nutritional state for difficult-to-sample populations, Life Sciences Research Office. *Journal of Nutrition* 120 (suppl): 1559–1600.

Anol, B. (2012) *Social Science Research: Principles, Methods, and Practices,* 2nd ed. Published under the Creative Commons Attibution-Non Commercial-ShareAlike 3.0 Unported License.

Atlanta Community Food Bank (2013) *Community Gardens: What is the Community Gardens Project?* Atlanta Community Food Bank. http://www.acfb.org/about/our-programs/community-gardens, accessed 14/10/2013.

AusAID (1999) Growing rice and protecting forests: an evaluation of three food production projects in S.E. Asia. Quality Assurance Series Number 15. June. Canberra: Commonwealth of Australia. ISBN 0 642 39992 1.

Bagamba, F., Bashaasha, B., Claessens, L. and Antle, J., (2012) Assessing climate change impacts and adaptation strategies for smallholder agricultural systems in Uganda. *Africa Crop Science Journal* 20, 303–316.

Barnett, R. (2001) Coping with the costs of primary care? Household and locational variations in the survival strategies of the urban poor. New Zealand: *Health and Place* 7(2):141–57. doi.org/10.1016/S1353–8292(01)00013–2

Barraclough, S. and P. Utting (1987) Food security trends and prospects in Latin America. Working Paper No. 99. Helen Kellogg Institute for International Studies, University of Notre Dame.

Berg, T. (1996) Dynamic management of Plant Genetic Resources: Potentials of Emerging Grass-Roots Movements. Study No. 1. Studies in Plant Genetic Resources. Plant Production and Protection Division, Food and Agriculture Organisation, Rome.

Berner, M. and K. O'Brien (2004) The shifting pattern of food security support: Food stamp and food bank usage in North Carolina. *Nonprofit and Voluntary Sector Quarterly* 33:655–72. doi: 10.1177/0899764004269145

Borlaug, N.E., O. Aresvik, I. Narvaez and R.G. Anderson (1969) A Green Revolution Yields a Golden Harvest. *Columbia Journal of World Business* 4: 9–19, September–October.

Brown, L.R. (1970) *Seeds of Change: The Green Revolution and Development in the 1970's.* New York: Praeger.

Bryman, A. (2008) *Social Research Methods.* 3rd ed. New York: Oxford University Press.

Caritas (2004) Village Food Security in Rural Laos. Issue No. 14. OneWorld Partnership.

Carter, I. (2001) *Community Grain Banks.* Excerpted from Improving Food Security, Village Volunteers, Tearfund.

Catholic Relief Services (CSR) (1997) Local capacity development: Principles and standards. Catholic Relief Services Discussion Paper No. 9, Baltimore.

Center for International Governance (CIGI) (2007) International Risk Report. Center for International Governance, Waterloo, Canada.

Chalmers Community Services Center (2014) Home Page. http://www.chalmerscenter.ca, accessed 16/05/2014.

Chambers, R. (1997) *Whose reality counts? Putting the first last.* London: Intermediate Technology Publications.

CIDA (1989) *Food Security: A working Paper for the 4A's.* Canadian, Area Coordination Group

Cohen, D.C. (2006) *Qualitative Research Guidelines Project.* July. http://www.qualres.org/HomeStra-3813.html, accessed 27/08/2013.

Community Food Bank of Southern Arizona (2009) Programs and Services: Farm/Gardens and Farmers' Markets, Community Food Bank. http://communityfoodbank.com/, accessed 02/11/2013.

Conor, W. (2011) *Lesson Learn Messages from the Field, Topic: Promoting food security through Rice Banks.* Dublin: Concern Worldwide.

Cooperative Societies Act (2003) *The Savings and Credit Cooperative Societies Regulations 2004.* Republic of Tanzania.

Crack, S., S. Turner and B. Heenan (2007) The changed face of voluntary welfare provision in New Zealand. *Health and Place* 13:188–204. doi.org/10.1016/j.healthplace.2005.12.001.

Curtis, M. (2013) Powering Up Smallholder Farmers to make Food Fair: A Five Point Agenda. Fairtrade Foundation: United Kingdom.

d'Silva, R. (2014) http://www.academia.edu/5460974/Ending_Hunger_and_Malnutrition_in_Rural_Context_through_Community_centered_food_security_systems, accessed 20/05/2014.

Dahlberg, A.C. and S.B. Trygger (2009) Indigenous Medicine and Primary Healthcare: The Importance of Lay Knowledge and Use of Medicinal Plants in Rural South Africa. *Human Ecology* 37:79–94. doi: 10.1007/s10745-009-9217-6

Daily Bread Food Bank (2005) *Who's Hungry: 2005 Profile of Hunger in the Greater Toronto Area.* Toronto: Daily Bread Food Bank.

Daily Monitor (2012) Farmers worried as caterpillars eat crops. *Daily Monitor*, Kampala. 14 June. http://www.monitor.co.ug/News/National/-/688334/1379580/-/awmklqz/-/index.html, accessed 16/05/2014.

Daly, G. (1996) Homeless: Policies, Strategies, and Lives on the Street. *Journal of Social Policy* 26:397–423. doi.org/10.1017/S0047279497345073

Datta. D. (2007) Community-managed rice banks: Lessons from Laos. *Development in Practice* 17:410–14. doi: 10.1080/09614520701337046

Davies, S. (1993) Are coping strategies a cop out? In J. Swift (ed.) New approaches to famine. IDS Bulletin No. 24:60–72.

Davies, S. (1996) *Adaptable Livelihoods: Coping with food insecurity in the Malian Sahel*, Oxford: Macmillan.

de Waal, A. (1991) Emergency food security in Western Sudan: What is it for? In S. Maxwell (ed.) *To Cure all Hunger: Food Policy and Food Security in Sudan.* London: Intermediate Technology.

Department of Agriculture, Forestry and Fisheries (DAFF) (2012) A framework for the development of smallholder farmers through cooperative development. Department of Agriculture, Forestry and Fisheries, Republic of South Africa.

Derrickson, J.P., A.G. Fisher and J.E.L. Anderson (2000) The Core Food Security Module Scale Measure Is Valid and Reliable When Used with Asians and Pacific Islanders. *American Society for Nutritional Sciences*130:2666–74

Development Fund, Norway (2011) Banking for the future: Savings, security and seeds. A short study of community seed banks in Bangladesh, Costa Rica, Ethiopia, Honduras, India, Nepal, Thailand, Zambia and Zimbabwe. The Development Fund/ Utviklingsfondet, Norway.

Devereux, S. and S. Maxwell (2001) Food security in Sub-Saharan Africa. Institute of Development Studies, United Kingdom.

Dinar, A. (2007) *Climate Change: The Final Blow for Agriculture in Africa?* World Bank Research Brief. August. Washington, DC.

East Africa Agribusiness 2014. Uganda's Agricultural season is changing over the years. March. http://ea-agribusiness.co.ug/tag/science/, accessed 16/05/2014.

Edakkandi, M.R. (2013) Community grain banks and food security of the tribal poor in India. *Development in Practice* 23:920–33.doi.org/10.1080/09614524.2013.811 469

Eicher, C. (1995) Zimbabwe Maize-based Green Revolution – Preconditions for Replication. *World Development* 23(5): 805–18. doi.org/10.1016/0305-750X(95)93983-R

Eide, W.B. (1990) Proceedings of the Agriculture-Nutrition Linkage Workshop. Vol. 1. February. Virginia.

Ellis, F. (2000) *Rural Livelihoods and Diversity in Developing Countries.* Oxford: Oxford University Press.

EPA (n.d.) What is Sustainability? What is EPA doing? How can I Help? United States Environment Protection Agency. http://www.epa.gov/sustainability/basicinfo.htm, accessed 28/08/2013.

Falcon, W.P., C.T. Kurien, F. Monckeberg, A.P. Okeyo, S.O. Olayide, F. Rabar and W. Tims (1987) The world food and hunger problem: Changing perspectives and possibilities, 1974–1984. In J.L. Gittinger and C. Hoisington (eds) *Food Policy: Integrating Supply, Distribution and Consumption.* Balitmore and London: Johns Hopkins University Press.

FAO and WFP (2009) The state of food insecurity in the world: Economic crises – impact and lessons learned. Food and Agriculture Organisation, Rome.

FAO (1983) World food security: A reappraisal of the concepts and approaches, Director-General's Report. Food and Agriculture Organisation, Rome.

FAO (1995) Sorghum and millets in human nutrition. FAO Food and Nutrition Series No. 27. Food and Agriculture Organisation, Rome.

FAO (1996) Rome Declaration on World Food Security and World Food Summit Plan of Action. Food and Agriculture Organisation Corporate Document Repository. http://www.fao.org/docrep/003/W3613E/W3613E00.HTM, accessed 07/10/2013.

FAO (2003) World Agriculture: Towards 2015/2030, Chapter 13. Climate Change and Agriculture – physical and human dimensions. Food Agriculture Organisation, Rome.

FAO (2004) A participative approach to food security. Food and Agriculture Organisation, Rome.

FAO (2006) Food Security. Policy Brief, Issue 2. June. Food and Agriculture Organisation, Rome.

FAO (2009) How to Feed the World in 2050. Food and Agriculture Organisation, Rome.

FAO (2011a) Global Food Losses and Food Waste: Extent, Causes and Prevention. Food and Agriculture Organisation, Rome.

FAO (2011b) The State of Food and Agriculture: Women in Agriculture: Closing the Gender Gap for Development. Food and Agriculture Organisation, Rome.

FAO (2014a) 38 countries meet anti-hunger targets for 2015: FAO lauds Millennium Development Goal, World Food Summit achievers. Food and Agriculture Organisation news article. 12 June. http://www.fao.org/news/story/en/item/177728/icode/, accessed 27/04/2014.

FAO (2014b) Save Food: Global initiatives on food loses and waste reductions: Key Findings. Food and Agriculture Organisation, Rome.

FAO (2014c) Joint FAO/WHO Second International Conference on Nutrition (ICN2), Regional Conference for Europe, Twenty-Ninth session. Bucharest, 2–4 April. Food and Agriculture Organisation, Rome.

FAO, IFAD and WFP (2012) The State of Food Insecurity in the World 2012. Economic growth is necessary but not sufficient to accelerate reduction of hunger and malnutrition. Food and Agriculture Organisation, Rome.

FAO, IFAD and WFP (2013) The State of Food Insecurity in the World: The multiple dimension of food security. Food and Agriculture Organisation, Rome.

FEWSNET (2014) Uganda: Food Security Outlook Update. March to June first season rains started on time in bimodal areas. March. Famine Early Warning Systems Network. http://www.fews.net/east-africa/uganda/food-security-outlook-update/mon-2014-03-3, accessed 29/04/2014.

Fisher, N.M. (1996) The potential of grain and forage legumes in mixed farming systems. In D. Younie (ed.) *Legumes in Sustainable Farming Systems.* British Grassland Society Occasional Symposium No. 30. Aberdeen, 2–4 September. pp. 290–9. 1 86320 420 2.

Fitzgerald, K. and J. Cameron (1989) Voluntary Organisations in Christchurch: An Overview of Staffing, Funding and Service (phase 1) Christchurch District Council of Social Services.

Flewwelling, P. and G. Hosch (2003) Review of the state of world marine capture fisheries management: Indian Ocean, Country review: India (East coast). Food and Agriculture Organisation, Rome.

Food Bank of Delaware (2011) Programs, Delaware: Food bank of Delaware, www.fbd.org/program/, accessed 25/05/2014.

Food Bank of South Africa (2013) Programmes: Community Development. http://www.foodbank.org.za, accessed 02/11/2013.

Food Bank of South Jersey (2014) We need your help to feed kids healthy lunches and snacks this summer. http://www.foodbanksj.org, accessed 16/05/2014.

Frankenberger, T.R. and D.M. Goldstein (1991) The Long and Short of it: Household Food Security: Coping Strategies and Environmental Degradation in Africa. Mimeo, Office of Arid Land Studies, University of Arizona.

Ghauri, P.N. and K. Grønhaug (2010) *Research methods in business studies: A Practical Guide.* (4th ed.). FT-Pearson.

Ghosh, P.K. (2004) Growth, yield, competition and economics of groundnut/cereal fodder intercropping systems in the semi-arid tropics of India. *Field Crops Research* 88(2–3):227–37. doi.org/10.1016/j.fcr.2004.01.015

Gittinger, J., N.R. Chernick and K. Saiter (1990) Household food security and the role of women. World Bank Discussion Paper 96. Washington DC: World Bank.

Gillespie, S. and J. Mason (1991) Nutrition-relevant actions: Some experiences from the eighties and lessons for the nineties. Nutrition Policy Discussion Paper No.10. Administrative Committee on Coordination/Subcommittee on Nutrition, United Nations, Geneva.

Global Environment Outlook – GEO5 (2012) Summary for Policy Makers, Chapter 16: Scenarios and Sustainability Transformation. United Nations Environment Programme (UNEP)

Global Food Bank Network (n.d.) What is food banking? Global Food Bank Network. http://www.foodbanking.org/site/PageServer?pagename=foodbanking_main, accessed 07/10/2013.

Global Hunger Project and Affiliates (2012) Consolidated financial report. December. Hunger Project, New York.

Hamilton, W.L., J.T. Cook, W.W. Thompson, L.F. Buron, E.A. Frongillo Jr., C.M. Olson and C.A. Wehler (1997) Household Food Security in the United States in 1995: Technical Report of the Food Security Measurement Project. Report prepared for the USDA, Food Consumer Service, Alexandria VA.

Hamoudi, A. (2010) Risk Aversion, Household Partition, and Consumption Smoothing in Rural Mexico, Verona University.

Hannah, S. and E. Dubey (n.d.) Global Sustainability Targets: Building Accountability for the 21st Century Earth Summit 2012. Stakeholder Forum.

Hayes, D. and R. Laudan (2009) *Food and Nutrition Encyclopedia*. New York: Marshall Cavendish.

Heald, C. and M. Lipton (1984) African food strategies and EEC's role: An interim assessment. IDS Commissioned Study No. 6. Institute of Development Studies, Brighton.

Heim, S., J. Stang and M. Ireland (2009) A Garden Pilot Project Enhances Fruit and Vegetable Consumption among Children. *Journal of the American Dietetic Association* 109: 1220–26.

Herath, G. and S. Jayasuriya (1996) Adoption of HYV Technologies in Asian Countries. *Asian Survey* 36(12):1184–1200. 36,903,181.

High Level Panel of Experts (2012) Climate Change and Food Security: A report by the High Level Panel of Experts on Food Security and Nutrition of the Committee on World Food Security. Food and Agriculture Organisation, Rome.

Hunger Project (THP) (2013a) Issues: World Hunger. http://www.thp.org/learn_more/issues/world_hunger, accessed on: 25.04.13.

Hunger Project (2013b) Invitation to apply for the position of president and chief executive officer. Hunger Project, New York.

Husbands, W. (1999) Food banks as Anti-hunger Organisations. In M. Koc, R. MacRae, L.A.J. Mougeot, and J. Welsh (eds) *For hungry cities: Sustainable urban food systems*. Ottawa: International Development Research Centre.

IFAD (2001) Rural Poverty Report 2001: The Challenge of Ending Rural Poverty. International Fund for Agricultural Development, Rome.

IFAD (2013) Smallholders, Food Security and the Environment. International Fund for Agricultural Development, Rome.

IFAD (2014) Food security: A conceptual framework. International Fund for Agricultural Development, Rome.

Inter Pares (2004) Community Based Food Security System: Local Solutions for Ending Chronic Hunger and Promoting Rural Development. Inter Pares Occasional Paper Series No. 4. Inter Pares, Ottawa.

IPC (2013) Report of the Integrated Food Security Phase Classification (IPC) Analysis for Uganda. Prepared by Uganda IPC Technical Working Group. November. IPC, Kampala.

IPCC (2007) Climate change 2007: Impacts, Adaptation and Vulnerability, Working Group II Contributions to the Fourth Assessment Report of the Intergovernmental Panel on Climate Change. Intergovernmental Panel on Climate Change.

IPCC (2014) Climate Change 2014: Impacts, Adaptation, and Vulnerability. Summary for Policy Makers. Intergovernmental Panel on Climate Change.

Iqbal, J., J.A. Cheema and M. An (2007) Intercropping of field crops in cotton for management of purple nutsedge (*Cyperus rotundus* L.). *Pakistan Journal of Botany* 40: 2383–91. doi: 10.1007/s11104-007-9400-8

Jensen E.S. (2002) The contribution of grain legumes, currently under-utilised in the EU, to a more environmentally-friendly and sustainable European LINK. Grain legumes for Sustainable Agriculture, Strasbourg, 26 September. Org Print 1852.

Jonsson, U. and D. Toole (1991) Household food security and nutrition: A conceptual analysis. Mimeo. UNICEF, New York.

Kaplan, A. (1999) The Developing of Capacity. Community Development Resource Association, Cape Town. Originally published as a development dossier by the United Nations Non-Governmental Liaison Service.

Ken, W. (2013) A better way to feed the World. Christensen Fund, San Francisco. http://www.christensenfund.org/2013/01/15/a-better-way-to-feed-the-world/, accessed 30.08.2013.

Kennes, W. (1990) The European Community and food security. IDS Bulletin 25. Institute of Development Studies, Brighton.

Kessy, F., O. Mashindano, A. Shepherd, A. and L. Scott (eds) (2013) *Translating Growth into Poverty Reduction, Beyond the Numbers*. Dar-es-Salaam: Mkuki Na Nyota Publishers.

Koc, M., R. MacRae, L.A.J. Mougeot and J. Welsh (eds) (1999) *Introduction: Food Security is a Global Concern*. In M. Koc, R. MacRae, L.A.J. Mougeot, and J. Welsh (eds) *For hungry cities: Sustainable urban food systems.* Ottawa: International Development Research Centre.

LaFond, A. and L. Brown (2003) A Guide to Monitoring and Evaluation of Capacity-Building: Interventions in the Health Sector in Developing Countries. MEASURE Evaluation Manual Series, Carolina Population Center, UNC, Chapel Hill.

Leathers, H. and P. Foster. (2009) *The world food problem: Toward ending under-nutrition in the third world.* Boulder CO: Lynne Rienner.

Lipton, M. (1989) *New Seeds and Poor People*. London: Unwin.

Lipton, M. (2005) Crop Science, poverty and family farm in a globalizing world. International Food Policy Research Institute Policy Brief 74. June.

Ludi, E. (2009) Climate Change, Water and Food Security. Background Note. March. ODI, London.

Malik, M.S., J.K Norsworthy, A.S. Culpepper, M.B. Riley and W. Bridges (2008) Use of wild radish (*Raphanus raphanistrum*) and rye cover crops for weed suppression in sweet corn. *Weed Science* 56:588–95. doi.org/10.1614/WS-08-002.1

Maxwell, S. (1988) National Food Security Planning: First Thoughts from Sudan. Paper presented to workshop on Food Security in Sudan. Institute of Development Studies, University of Sussex. 3–5 October 1988.

Mbowa, S., J. Mawejje and I. Kasiry, (2012) Understanding the recent food price trends in Uganda. Economic Policy Research Centre Fact Sheet No. 4. Kampala.

Mellor, J. (1990) Global food balances and food security, in C.K. Eicher and J.M. Staatz (eds) *Agriculture Development in the Third World*. 2nd ed. Baltimore: Johns Hopkins University Press.

Merriam, S.B. (2009) *Qualitative research: A guide to design and implementation*. San Francisco: John Wiley and Jossey-Bass.

Michael, D. (1993) *Rooms in the House of Stone: The "Thistle" Series of Essays*. Minneapolis: Milkweed editions. June.

Ministry of Agriculture, Animal Industry and Fisheries (2001) *National Agricultural Advisory Services Act*. Kampala.

Ministry of Agriculture, Animal Industry and Fisheries (MAAIF) and Ministry of Health (MoH) (2003) Uganda Food and Nutrition Policy (UFNP). Kampala.

Ministry of Finance, Planning and Economic Development (MoFPED) (2004) Poverty Eradication Action Plan (PEAP) 2004–2009. Kampala.

Ministry of Finance Planning and Economic Development (2005) Poverty Eradication Action Plan 2004/05–2007/08 – A Summary Version. Kampala.

Ministry of Health, Health Systems 20/20, and Makerere University School of Public Health, (2012) *Uganda Health System Assessment 2011*. Kampala, Uganda and Bethesda MD: Health Systems 20/20 project, Abt Associates Inc.

Ministry of Water, Lands, and Environment (2002) The National Forest Plan. Kampala.

Moldofsky, Z. (2000) Meals Made Easy: A Group Program at a Food Bank. *Social Work with Groups* 23:83–96. doi: 10.1300/J009v23n01_06

Natural Resource Defense Council (NRDC) (2010) *Eat Green: Everyday food choices affect global warming and the environment, Natural Resource Defense Council*. http://www.nrdc.org/globalwarming/files/eatgreenfs_feb2010.pdf, accessed 10/11/2013.

Neely, C. and A. Fynn (n.d.) Critical Choices for Crop and Livestock Production systems that enhance productivity and builds ecosystem resilience. SOLAW Background Thematic Report TR-11. Food and Agriculture Organisation, Rome.

Neuman, W.L. (2007) *Basics of Social Research: Quantitative and Qualitative Approaches*. 2nd ed. Boston: Allyn and Bacon.

New Zealand Council of Christian Social Services (NZCCSS) (2005) Forgotten Poverty? Poverty Indicator Project: Food Bank Study Final Report. Wellington, New Zealand.

Nichols-Casebolt, A. and P.M. Morris (2001) Making ends meet: Private food assistance and the working poor. *Journal of Social Service Research* 28:2–22. doi.org/10.1300/J079v28n04_01

Nobel Peace Prize (1970) The Nobel Peace Prize: Norman Borlaug Biographical. Nobel Foundation. http://www.nobelprize.org/nobel_prizes/peace/laureates/1970/borlaug-bio.html, accessed 25/04/2014

Ntabadde, C.M. (2013) Kasese floods: Affected population now at 25,445. The Uganda Red Cross Society. 9 May. http://www.redcrossug.org/component/content/article/3-newsflash/433-kasese-floods-affected-population-now-at-25-445.html, accessed 16/05/2014

Nyeko, B. (compiler) (1996) *Uganda*. Santa Barbara CA: Clio Press.

Odero, K. (2011) The role of traditional, local and indigenous knowledge in responding to climate change: Local global perspectives. Presented at the Climate Change Symposium 2011, Africa Adapt.

Oshaug, A. (1985) The composite concept of food security. In W.B. Eide, G. Holmboe-Ottesen, A. Oshaug, D. Perer, S. Tilakaratna and M. Wandel (eds) Introducing nutritional considerations into rural development programmes with focus on agriculture: a theoretical contribution. Development of Methodology for the Evaluation of Nutritional Impact of Development Programme Report No. 1. Institute of Nutrition Reseach, University of Oslo.

Osmani, S. (1995) The entitlement approach to famine: An assessment. In K. Basu, P. Pattanaik and K. Suzumura (eds) *Choice, Welfare and Development*. Oxford; Oxford University Press.

Oya, K (2001) Capacity Building for Community Based Resource Management: Lessons from Southeast Asian countries. United Nations Centre for Regional Development, Nagoya, Japan.

Philips, T. and D. Taylor (1990) Optimal control of food insecurity: A conceptual framework. *American Journal of Agriculture Economics* 74: 1304–10.

Piper, H. and H. Simons (2005) Ethical responsibility in social research. In B. Somekh and C. Lewin (eds) *Research Methods in the Social Sciences*. London: Sage.

Proteau L. and F. Wolff (2008) On the relational motive for volunteer work. *Journal of Economic Psychology* 29:314–35. doi.org/10.1016/j.joep.2007.08.001

Rasch, G. (1966) An item analysis that takes individual differences into account. Br. J. Math. Stat. Psych. 4:321–33.

Ravinder, R.C., V.A Tonapi, P.G. Bezkorowajnyj, S.S. Navi and N. Seetharama (2007) Seed System Innovations in the semi-arid tropics of Andhra Pradesh. International Livestock Research Institute (ILRI), Patancheru, Andhra Pradesh.

Ray, D.K., N.D. Mueller, P.C. West and J.A. Foley (2013) Yield Trends Are Insufficient to Double Global Crop Production by 2050. *PLoS ONE* 8(6):e66428. doi:10.1371/journal.pone.0066428

Reardon, T. and Malton, P. (1989) Seasonal food insecurity and vulnerability in drought-affected regions of Burkina-Faso. In D.E. Sahn (ed.) *Seasonal Variability in Third World Agriculture: The Consequences for Food Security*. Baltimore and London: Johns Hopkins University Press.

Reddy, T. and B. Adolph (2002) Report of a visit to AGRAGAMEE: Grain banks for food security in tribal areas of Orissa. Department for International Development (DFID), UK.

REseau de LUtte contre la FAim (RELUFA) (2008) Community Grain Banks. http://www.relufa.org/programs/foodsovereignty/communitycerealbanks.htm, accessed 21/05/2014.

Republic of Uganda (1995) *Constitution of the Republic of Uganda*. Constituent Assembly of the Republic of Uganda, Kampala.

Republic of Uganda (2013) The State of Uganda Population Report 2013. Theme: Population and Social Transformation: Addressing the needs of Special Interest Groups. Kampala.

Reutilinger, S. and K. Knapp (1980) Food security in food deficit countries. World Bank Staff Working Paper 393. World Bank, Washington DC.

Reutilinger, S. (1982) Policies for food security in food-importing developing countries. In A.H. Chisholm and R. Tyers (eds) *Food Security: Theory, Policy, and Perspectives from Asia and Pacific Rim.* Massachusetts: Lexington Books.

Reutilinger, S. (1985) Food security and poverty on LDCs. *Finance and Development* 22: 7–11.

Riches, G. (1986) Food banks and the Welfare Crisis. Ottawa: Canadian Council on Social Development.

Riches, G. (2002) Food Banks and Social Security: Welfare Reform, Human Rights and Social Policy. Lessons from Canada? *Social Policy and Administration* 36:648–63. doi: 10.1111/1467-9515.00309.

Rosegrant, M., X. Cai, S. Cline and N. Nakagawa (2002) The Role of Rain-fed Agriculture in the Future of Global Food Production. EPTD Discussion Paper 90. Environment and Production Technology Division, International Food Policy Research Institute, Washington DC.

Sah, D.E. (1989) A conceptual framework for examining the seasonal aspects of household food insecurity. In D.E. Sahn (ed.) *Seasonal Variability in Third World Agriculture: The Consequences for Food Security.* Baltimore and London: Johns Hopkins University Press.

Saris, A.H. (1989) Food security and international security. Discussion Paper 302. Centre for Economic Policy Research, Washington DC.

Satheesh, P.V. (n.d.) Food Security for Dryland Communities. Director, Deccan Development Society. http://ddsindia.com/www/foodsec_dryland.htm, accessed 20/05/204.

Scoones, I. (1998) Sustainable rural livelihoods: A framework for analysis, IDS Working Paper 72. Brighton: IDS.

Shah, A. (2008) Global Food Crisis 2008. *Global Issues,* 10 August. http://www.globolissues.org/article/758/global-food-crisis-2008, accessed 21.03.2014.

Shenggen, F. (2011) G20 Ministers of Agriculture Must Focus on Smallholder Farmers to Achieve Food Security and Prevent Food Price Volatility. Press statement, Director General, International Food Policy Research Institute (IFPRI). 15 June.

Siamwalla, A. and A. Valdes (1980) Food insecurity in developing countries. *Food Policy* 5(4): 258–72.

Soil Association (1997) *A Share in the Harvest: A feasibility study for Community Supported Agriculture A Participatory Approach toward Sustainable Agriculture.* Bristol: Soil Association.

Sperling, L., T. Remington and J.M. Haugen (2006) Seed Aid for Seed Security: Advice for Practitioners. Practice Briefs 1–10. International Centre for Tropical Agriculture and Catholic Relief Services, Rome.

Staatz, J. (1990) Food security and agriculture policy: Summary. Proceedings of the Agriculture-Nutrition Linkage Workshop Vol. 1. Virginia, February.

Starkey, L.J. and H.V. Kuhnlein (1996) Nutrient adequacy of urban food assistance provisions. *Journal of the American Dietetic Association* 96(9):A-55. doi. org/10.1016/S0002-8223(96)00495-6

Starkey, L.J., K. Gray-Donald and H.V. Kuhnlein (1999) Nutrient intake of food bank users is related to frequency of food bank use, household size, smoking, education and country of Birth. *Journal of Nutrition* 129: 883–89.

Starkey, L.J., H.V. Kuhnlein and K. Gray-Donald (1998) Food bank users: Sociodemographic and nutritional characteristics. *Canadian Medical Association Journal* 158(9): 1143–49.

Stephens, R., P. Frater and C. Waldegrave (2000) *Below the Line: An Analysis of Income Poverty in New Zealand, 1984–1998.* Wellington: Victoria University of Wellington.

Swaminatnan, M.S. (n.d.) Grain Seed Bank. M.S. Swaminathan Research Foundation. Jeypore, Odisha. http://www.mssrf.org/bd/jeypore/Grain-bank.pdf, accessed 21/05/2014.

Tarasuk, V. and H. MacLean (1990) The food problems of low-income single mothers: An ethnographic study. *Caandian Home Economics Journal* 40:7682.

Tarasuk, V. (2005) Household Food Insecurity in Canada. *Topics in Clinical Nutrition* 20:299–312.

Teron, A.C. and V. Tarasuk (1999) Charitable Food Assistance: What are Food Bank users Receiving? *Canadian Journal of Public health* 90(6):282–304.

The Hunger Project (1985). *Ending Hunger: An Idea Whose Time Has Come?* Praeger Press, New York.

The Hunger Project Uganda (THP-Uganda) (2012) *Annual Report 2012.*

Thériault, L. and L. Yadlowski (2000) Revisiting the Food Bank Issues in Canada. *Canadian Social Work Review* 17: 205–23.

Timmer, C.P., W.P. Falcon and R.S. Pearson (1983) *Food Policy Analysis.* Published for the World Bank by the Johns Hopkins University Press, Baltimore and London.

Uganda (2005) *Uganda districts information handbook*. Kampala: Fountain.

Uganda Bureau of Statistics (UBOS) (2002) Uganda Population and Housing Census. Uganda Bureau of Statistics, Kampala.

Uganda Bureau of Statistics (2010) Uganda National Household Survey 2009/2010, Socio-economic module, abridged report. November. UBOS, Kampala.

Uganda Bureau of Statistics (2006) Uganda National Household Survey 2005/2006, Report on the Socio-economic module. December. UBOS, Kampala

Uganda Savings and Credit Cooperation Union (2014) Welcome to UCSCU. Kampala. http://www.ucscu.co.ug/#, accessed 15/05/2014.

UN (1987) Report of the World Food Conference New York, 5–16 November 1974. Rome.

UN (1988) Towards sustainable food security: Critical issues. Report by the secretariat, World Food Council, Fourteenth Ministerial Session, 23–26 May. Nicosia, Cyprus.

UN (1997) UN Conference on Environment and Development (1992). United Nations 23 May 1997. http://www.un.org/geninfo/bp/enviro.html, accessed 25/08/2013.

UN (1998) Kyoto Protocol to the United Nations Framework Convention on Climate Change. United Nations.

UN (2013) World Economic and Social Survey 2013: Sustainable Development Challenge. United Nations Department of Economics and Social Affairs.

UN Rio+20 (2013) What is Sustainability? United Nations Rio+20.

UNBconnect (2014) Rice banks help CHT communities meet food shortage. Reported by **UNBconnect**, on 21 May, 06:05:43 am. http://unbconnect.com/rice-bank-2/#&panel1-5, accessed 21/05/2014.

UNDP (2008) Fighting Climate Change – Human Solidarity in a Divided World. UNDP, New York.

UNEP (2014) A changing climate creates pervasive risks but opportunities exist for effective responses. Intergovernmental Panel on Climate Change Report. 31 March. Yokohama, Japan. United Nations Environment Programme. http://www.unep.org/newscentre/Default.aspx?DocumentID=2764andArticleID=10773, accessed 01/04/2014.

UNEP (2011) Ready, Willing and Able: Empowering countries to meet the climate change challenge. United Nations Environment Programme

UNESCO (2013) Education for all global monitoring report: Regional fact sheet: Education in Eastern Africa. January. ED/EFS/MRT/2013/FS/1 REV.2, UNESCO, France.

UNICEF (1990) Strategy of improved nutrition of children and women in developing countries: A UNICEF Policy Review. New York: UNICEF.

Union of Concerned Scientists (2013) Food and Agriculture. http://www.ucsusa.org/food_and_agriculture/our-failing-food-system/industrial-agriculture/, accessed 30/08/2013.

United Nations Economic and Social Council /Economic Commission for Africa (2012) Status of Food Security in Africa. Eighth Session of the Committee on Food Security and Sustainable Development and Regional Implementation Meeting for the Twentieth Session of the Commission on Sustainable Development. Addis Ababa, 19–21 November.

United Nations Sustainable Development Knowledge Platform (2011) Commission on Sustainable Development (CSD): *Future we want – Outcome document.* United Nations Division for Sustainable Development (UN-DESA). http://sustainabledevelopment.un.org/csd.html, accessed 28/08/2013.

Uttley, S. (1997) Hunger in New Zealand: A Question of Rights? In G. Riches (ed.) *First World Hunger: Food Security and Welfare Politics.* New York: St Martin's Press.

Valdes, A. and P. Konandreas (1981) Assessing food insecurity based on national aggregates in developing countries. In A. Valdes (ed.) *Food security in developing countries.* Boulder CO: Westview.

Vanloqueren, G. and V.P. Baret (2009) How agriculture research systems shape a technological regime that develops genetic engineering but locks out agro-ecological innovations. *Research Policy* 38(6):971–83. doi.org/10.1016/j.respol.2009.02.008

Village Earth (2013) Community garden providing food to widows and vulnerable children. Village Earth, Namibia.

Von Braun, J. (1991) Policy agenda for famine prevention in Africa. Food Policy Report, International Food Policy Research Institute, Washington DC.

Weber, M.T. and T.S. Jayne (1991) Food security and its relationship to technology, institutions, policies and human capital. In L.J. Glenn and J.T. Bonnen (eds) *Social Science Agriculture Agendas and Strategies.* East Lansing: Michigan State University.

Weine, S. (2013) Incorporating multi-methodological approaches into prevention research with refugees and migrants. Paper presented at the International Congress on Qualitative Inquiry, Urbana-Champaign. May.

World Health Organisation (WHO) (2012) Global and regional food consumption patterns and trends. April.

Wiggins, S. (2008) Rising Food Prices – A Global Crisis. ODI Briefing Paper No. 37. London.

Wolfe, W.S., C.M. Olson, A. Kendall and E.A. Frongillo Jr. (1996) Understanding food insecurity in the elderly: A conceptual framework. *Journal of Nutritional Education* 28(2): 92–100. doi.org/10.1016/S0022-3182(96)70034-1

World Bank (1986a) Poverty and Hunger: Issues and options for Food Security in Developing Countries. World Bank Policy Study. International Bank for Reconstruction and Development and World Bank, Washington DC.

World Bank (2013) Turn down the heat, Climate extremes, regional impacts and case for resilience. International Bank for Reconstruction and Development and World Bank, Washington DC.

World Bank (2013a) Agriculture and Rural Development: Land and Food Security. International Bank for Reconstruction and Development and World Bank, Washington DC. http://web.worldbank.org/WBSITE/EXTERNAL/TOPICS/EXTARD/0,,contentMDK:23284610~pagePK:148956~piPK:216618~theSitePK:336682,00.html, accessed 09.09.2013

World Bank (2013b) Africa's Food Markets Could Create One Trillion Dollar Opportunity by 2030. Press Release, 4 March. International Bank for Reconstruction and Development and World Bank, Washington DC.

World Commission on Environment and Development (WCED) (1987) *Our common future.* Oxford: Oxford University Press.

World Food Programme (2008) FAO/ WFP assessment of the impact of 2007 floods on food and agriculture in eastern and northern Uganda. Special report. FAO Global Information and Early Warning System on Food and Agriculture. World Food Programme.

World Food Programme (2009) World Food Program Says hunger kills more that AIDS, malaria, tuberculosis combined. World Food Programme release to Xinhua, Thursday 4 June.

Yameogo, S.P. (2013) Community Grain Banks, Tearfund International Learning Zone. http://tilz.tearfund.org/en/resources/publications/footsteps/footsteps_31-40/footsteps_32/community_grain_banks/, accessed 21/05/2014.

Zipperer, S. (1987) *Food Security and Agriculture Policy and Hunger.* Harare: Foundation for Education with Production.

CURRENT AFRICAN ISSUES PUBLISHED BY THE INSTITUTE

Recent issues in the series are available electronically
for download free of charge www.nai.uu.se

1981

1. *South Africa, the West and the Frontline States. Report from a Seminar.*
2. Maja Naur, *Social and Organisational Change in Libya.*
3. *Peasants and Agricultural Production in Africa. A Nordic Research Seminar. Follow-up Reports and Discussions.*

1985

4. Ray Bush & S. Kibble, *Destabilisation in Southern Africa, an Overview.*
5. Bertil Egerö, *Mozambique and the Southern African Struggle for Liberation.*

1986

6. Carol B.Thompson, *Regional Economic Polic under Crisis Condition. Southern African Development.*

1989

7. Inge Tvedten, *The War in Angola, Internal Conditions for Peace and Recovery.*
8. Patrick Wilmot, *Nigeria's Southern Africa Policy 1960–1988.*

1990

9. Jonathan Baker, *Perestroika for Ethiopia: In Search of the End of the Rainbow?*
10. Horace Campbell, *The Siege of Cuito Cuanavale.*

1991

11. Maria Bongartz, *The Civil War in Somalia. Its genesis and dynamics.*
12. Shadrack B.O. Gutto, *Human and People's Rights in Africa. Myths, Realities and Prospects.*
13. Said Chikhi, Algeria. *From Mass Rebellion to Workers' Protest.*
14. Bertil Odén, *Namibia's Economic Links to South Africa.*

1992

15. Cervenka Zdenek, *African National Congress Meets Eastern Europe. A Dialogue on Common Experiences.*

1993

16. Diallo Garba, *Mauritania–The Other Apartheid?*

1994

17. Zdenek Cervenka and Colin Legum, *Can National Dialogue Break the Power of Terror in Burundi?*
18. Erik Nordberg and Uno Winblad, *Urban Environmental Health and Hygiene in Sub-Saharan Africa.*

1996

19. Chris Dunton and Mai Palmberg, *Human Rights and Homosexuality in Southern Africa.*

1998

20. Georges Nzongola-Ntalaja, *From Zaire to the Democratic Republic of the Congo.*

1999

21. Filip Reyntjens, *Talking or Fighting? Political Evolution in Rwanda and Burundi, 1998–1999.*
22. Herbert Weiss, *War and Peace in the Democratic Republic of the Congo.*

2000

23. Filip Reyntjens, *Small States in an Unstable Region – Rwanda and Burundi, 1999–2000.*

2001

24. Filip Reyntjens, *Again at the Crossroads: Rwanda and Burundi, 2000–2001.*
25. Henning Melber, *The New African Initiative and the African Union. A Preliminary Assessment and Documentation.*

2003

26. Dahilon Yassin Mohamoda, *Nile Basin Cooperation. A Review of the Literature.*

2004

27. Henning Melber (ed.), *Media, Public Discourse and Political Contestation in Zimbabwe.*

28. Georges Nzongola-Ntalaja, *From Zaire to the Democratic Republic of the Congo.* (Second and Revised Edition)

2005

29. Henning Melber (ed.), *Trade, Development, Cooperation – What Future for Africa?*

30. Kaniye S.A. Ebeku, *The Succession of Faure Gnassingbe to the Togolese Presidency – An International Law Perspective.*

31. J.V. Lazarus, C. Christiansen, L. Rosendal Østergaard, L.A. Richey, Models for Life – Advancing antiretroviral therapy in sub-Saharan Africa.

2006

32. Charles Manga Fombad & Zein Kebonang, *AU, NEPAD and the APRM – Democratisation Efforts Explored.* (Ed. H. Melber.)

33. P.P. Leite, C. Olsson, M. Schöldtz, T. Shelley, P. Wrange, H. Corell and K. Scheele, *The Western Sahara Conflict – The Role of Natural Resources in Decolonization.* (Ed. Claes Olsson)

2007

34. Jassey, Katja and Stella Nyanzi, *How to Be a "Proper" Woman in the Times of HIV and AIDS.*

35. M. Lee, H. Melber, S. Naidu and I. Taylor, *China in Africa.* (Compiled by Henning Melber)

36. Nathaniel King, *Conflict as Integration. Youth Aspiration to Personhood in the Teleology of Sierra Leone's 'Senseless War'.*

2008

37. Aderanti Adepoju, *Migration in sub-Saharan Africa.*

38. Bo Malmberg, *Demography and the development potential of sub-Saharan Africa.*

39. Johan Holmberg, *Natural resources in sub-Saharan Africa: Assets and vulnerabilities.*

40. Arne Bigsten and Dick Durevall, *The African economy and its role in the world economy.*

41. Fantu Cheru, *Africa's development in the 21st century: Reshaping the research agenda.*

2009

42. Dan Kuwali, *Persuasive Prevention. Towards a Principle for Implementing Article 4(h) and R2P by the African Union.*

43. Daniel Volman, *China, India, Russia and the United States. The Scramble for African Oil and the Militarization of the Continent.*

2010

44. Mats Hårsmar, *Understanding Poverty in Africa? A Navigation through Disputed Concepts, Data and Terrains.*

2011

45. Sam Maghimbi, Razack B. Lokina and Mathew A. Senga, *The Agrarian Question in Tanzania? A State of the Art Paper.*

46. William Minter, *African Migration, Global Inequalities, and Human Rights. Connecting the Dots.*

47. Musa Abutudu and Dauda Garuba, *Natural Resource Governance and EITI Implementation in Nigeria.*

48. Ilda Lindell, *Transnational Activism Networks and Gendered Gatekeeping. Negotiating Gender in an African Association of Informal Workers.*

2012

49. Terje Oestigaard, *Water Scarcity and Food Security along the Nile. Politics, population increase and climate change.*

50. David Ross Olanya, *From Global Land Grabbing for Biofuels to Acquisitions of AfricanWater for Commercial Agriculture.*

2013

51. Gessesse Dessie, *Favouring a Demonised Plant. Khat and Ethiopian smallholder enterprise.*

52. Boima Tucker, *Musical Violence. Gangsta Rap and Politics in Sierra Leone.*

53. David Nilsson, *Sweden-Norway at the Berlin Conference 1884–85. History, national identity-making and Sweden's relations with Africa.*

54. Pamela K. Mbabazi, *The Oil Industry in Uganda; A Blessing in Disguise or an all Too Familiar Curse? Paper presented at the Claude Ake Memorial Lecture.*

55. Måns Fellesson & Paula Mählck, *Academics on the Move. Mobility and Institutional Change in the Swedish Development Support to Research Capacity Buildiing in Mozambique.*

56. Clementina Amankwaah. *Election-Related Violence: The Case of Ghana.*

57. Farida Mahgoub. *Current Status of Agriculture and Future Challenges in Sudan.*

58. Emy Lindberg. *Youth and the Labour Market in Liberia – on history, state structures and spheres of informalities.*

59. Marianna Wallin. *Resettled for Development. The Case of New Halfa Agricultural Scheme, Sudan.*

60. Joseph Watuleke. *The Role of Food Banks in Food Security in Uganda. The Case of the Hunger Project Food Bank, Mbale Epicentre.*

www.ingramcontent.com/pod-product-compliance
Lightning Source LLC
Chambersburg PA
CBHW080208300326
41934CB00039B/3414